PCR METHODS IN FOODS

PCR METHODS IN FOODS

Edited by
John Maurer
The University of Georgia, Athens
Athens, GA, USA

Dr. John Maurer
252 Poultry Diagnostic and Research Center
College of Veterinary Medicine
The University of Georgia
Athens, GA 30602
USA

ISBN   978-1-4419-3933-3   e-ISBN   978-0-387-31702-1

Printed on acid-free paper.

9 8 7 6 5 4 3 2 1

springeronline.com

Preface

This book will introduce non-molecular biologists to diagnostic PCR-based technologies for the detection of pathogens in foods. By the conclusion of this book, the reader should be able to: 1) understand the principles behind PCR including real-time; 2) know the basics involved in the design, optimization, and implementation of PCR in food microbiology lab setting; 3) interpret results; 4) know limitations and strengths of PCR; and 5) understand the basic principles behind a new fledgling technology, microarrays and its potential applications in food microbiology. This book will provide readers with the latest information on PCR and microarray based tests and their application towards the detection of bacterial, protozoal and viral pathogens in foods. Figures, charts, and tables will be used, where appropriate, to help illustrate concepts or provide the reader with useful information or resources as an important starting point in bringing molecular diagnostics into the food microbiology lab. This book is not designed to be a "cookbook" PCR manual with recipes and step-by-step instructions but rather serve as a primer or resource book for students, faculty, and other professionals interested in molecular biology and its integration into food safety.

Table of Contents

Preface v

Chapter 1. *PCR Basics*

    Amanda Fairchild, M.S., Margie D. Lee DVM, Ph.D.,
    and John J. Maurer, Ph.D. 1

Chapter 2. *The Mythology of PCR: A Warning to the Wise*

    John J. Maurer, Ph.D. 27

Chapter 3. *Sample Preparation for PCR*

    Margie D. Lee, DVM, Ph.D. and Amanda Fairchild, M.S. 41

Chapter 4. *Making PCR a Normal Routine of the Food Microbiology Lab*

    Susan Sanchez, Ph.D. 51

Chapter 5. *Molecular Detection of Foodborne Bacterial Pathogens*

    Azlin Mustapha, Ph.D. and Yong Li, Ph.D. 69

Chapter 6. *Molecular Approaches for the Detection of Foodborne Viral Pathogens*

    Doris H. D'Souza and Lee-Ann Jaykus 91

Chapter 7. *Molecular Tools for the Identification of Foodborne Parasites*

    Ynes Ortega, Ph.D. 119

Index 147

CHAPTER 1

PCR Basics

Amanda Fairchild[1], M.S., Margie D. Lee[1,2] DVM, Ph.D.,
and John J. Maurer[1,2*], Ph.D.

*Poultry Diagnostic & Research Center[1], College of Veterinary Medicine,
The University of Georgia, Athens, GA 30602*
*Center for Food Safety[2], College of Agriculture and Environmental Sciences,
The University of Georgia, Griffin, GA 30223*

Introduction
  Using Molecular Methods to Identify Microbial Pathogens
The Theory Behind PCR
Thermocycler Technology
Detection
Advanced PCR Technologies
  Real-Time PCR
  Multiplex PCR
  Terminal Restriction Fragment Length Polymorphisms
  Microarrays
Design and Optimization of Diagnostic PCR as Applicable to Food
Microbiology
  Systematic Approach to Creating Your Own PCR
Access DNA Databases to Retrieve Sequences or Search for DNA Matches
References

INTRODUCTION

The safety of your food supply is an important goal of the U.S. government and diagnostic food microbiologists across the country. Up to 5,000 deaths and 76 million illnesses in the U.S. each year are associated with the consumption of foods laced with pathogenic bacteria (53), costing the U.S. an estimated $6.5–$34.9 billion annually (8). Even though bacteria have been shown to be the cause of the majority of food-related illnesses, the government does not have a mechanism for detecting and accounting for the losses due to other common foodborne pathogens, such as viruses and protozoa. Detection, identification, and quantification of foodborne pathogens are often made difficult by the low numbers of pathogenic organisms and interference from the food matrix that is being sampled. Bacterial pathogens of particular importance include *Listeria, Campylobacter, Escherichia coli,* and *Salmonella* (53), and the norovirus and

*Corresponding author. Phone: (706) 542-5071; FAX: (706) 542-5630; e-mail: jmaurer@vet.uga.edu.

hepatitis A virus are currently regarded as important foodborne viruses (44). However, since the advent of the polymerase chain reaction, finding these few pathogenic microorganisms in otherwise innocent looking provisions is becoming easier, mainstreamed, and second nature to many diagnostic laboratories. The polymerase chain reaction (PCR) is a simple way to quickly amplify specific sequences of target DNA from indicator organisms to an amount that can be viewed by the human eye with a variety of detection devices. A goal of the present-day food microbiology research laboratory is to use the growing database of bacterial genomic information, made available by researchers mapping unique identifier genes of foodborne pathogens, to design monitoring systems capable of analyzing various incoming samples for hundreds of different organisms accurately and efficiently.

Using Molecular Methods to Identify Microbial Pathogens. Prior to the 1980s and the advent of PCR, identification of microbial pathogens relied on bacteriological methods to enrich and isolate the organism from clinical or food sample, and subsequent biochemical and/or immunological tests to confirm the microbe's identity. During the past 30 years, we have gained tremendous insights into how microorganisms spread and cause disease. In several instances, pathology associated with many bacterial illnesses is attributed to a single gene (4, 9, 18, 75, 97). Other pathogens like *Salmonella* are more complex, requiring coordinate regulation of several virulence gene sets to cause disease (49). Therefore, the organism's genetics or genotype dictates its ability to cause disease, or the severity of the illnesses associated with it. Many of these virulence genes are unique to the pathogen and subsequently make useful markers for identifying said pathogen (37, 41, 86, 95). With several bacterial genomes completed, we now know the genetic basis for phenotypes that have been useful markers for distinguishing pathogens from closely related commensals that inhabit the same niche. By identifying gene(s) associated with phenotype (e.g., O157 serotype), we have identified a marker with greater specificity than afforded by the actual antigen itself, especially when confronting cross-reactivity and false positive associated with the immunological test (50). Although quite specific, the early molecular-based method, DNA: DNA hybridization had limited utility due to its limited sensitivity, time length, or safety issues associated with the use of radioactive probes (77, 80). Even with the introduction of nonradiometric methods for detecting hybridization of probe to its target gene, there was still the limitation of sensitivity, [i.e., the ability to detect the fewest cells possible (80)]. *How could one amplify the target gene enough to detect its presence in the sample contaminated with organism X?*

THE THEORY BEHIND PCR

The concept of the PCR was first described by Panet and Khorana in 1974(64) and owes its name to Dr. Kary Mullis and colleagues, who developed the process over the course of 4 months in 1983 at the Cetus Corporation. While driving down Highway 128 in Mendocino County, California, Dr. Mullis let his mind slip back to the lab and a burning question that could not escape his mind. How could someone go about reading the sequence known as DNA, the language of our genes and blueprint for our existence? Like perfectly crafted ball and socket joints, the oligonucleotide base pairs within the DNA molecule bond to one another as the entire length of the ladder-shaped molecule twists into a corkscrew shape. Dr. Mullis referred to DNA structure as something like a mass of "unwound and tangled audio tape on the floor of the car in the dark" (58). In 1953, Drs. Watson and Crick had mentioned the biological significance of the DNA molecule with its complementary base pairing that suggested "a possible copying mechanism" for genetic material (91). One strand of DNA could be a template for the formation of a new complementary chain, and in the end you could have two DNA ladders, identical in every way. Dr. Mullis' development of PCR has extrapolated the copy machine theory one step further. He stated this simply as an analogy to a "'Find' sequence in a computer search" (58). This technique would have equivalent power to the latest computer displaying results of finding a document that consisted of just one word taking up 20 kilobytes of space on a hard drive the size of 150 Gigabytes littered with files of different types and sizes; like finding the code for blue eyes—and that code only—within the code that sums up every single trait for a person.

The second aspect Dr. Mullis had to account for in this process would be the ability of the chemical program to display the located sequence in a large enough fashion to be detected by the human eye. Dr. Mullis knew that if he could produce a short piece of DNA to find a sequence flanking a gene of interest and then start a process that could make the sequence reproduce itself over and over (hence a chain reaction), the concept of PCR could be realized. After all, it was already known that DNA innately makes a copy of itself when cells divide, so that each daughter cell can have a copy. If he introduced his "find" search strings, or primers, into a tube with DNA encouraged to uncoil from its natural double helix by heating and a biological glue (polymerase) to attach deoxynucleoside triphosphates (dNTPs) to the freshly uncoiled DNA at the location, where the primers bonded, a new copy of the DNA would be produced as specified by the primers that included the gene(s) of interest as the temperature decreased. The DNA polymerase that is used in some PCR reactions is made from the bacteria *Thermus aquaticus*, which was found originally in Yellowstone thermal water reservoirs. It is stable in temperatures above that which denatures DNA, making it a perfect enzyme for the job of attaching free dNTPs to make a new strand of DNA. There are other polymerases available that can actually proofread the addition of dNTPs, so there will be no errors made in the synthesis of longer PCR products. After each cycle in a PCR assay, the amount of DNA present doubles, so repeat the cycle and there are 4 copies of the gene, repeat again and there are 8 copies. With 30 cycles of this process, there would theoretically be just over a billion copies of the sequence in question ($2^{30} = 1.07$ billion). Find a way to tag each copy to make it visible to the human eye, with more copies making a stronger visible signal and you have proof of the presence of the small sequence embedded within the large DNA molecule you started with. A widely used method for viewing PCR products involves running them on an agarose gel, staining the gel with ethidium bromide, and observing and photographing the gel on ultraviolet (UV) light source (56). The process is relatively fast, dictated by the amount of time it takes to heat the DNA strands until they will separate, the time to reduce the temperature so the primers bind to the single-stranded DNA, and the time allowed for the polymerase to add individual deoxynucleoside triphosphates to extend the forming DNA molecule. Dr. Mullis stated that scientists claimed that PCR made DNA research boring (57). Even though PCR is often considered "cookbook chemistry" because of its simplicity, his suggestion could not be further from the truth. For example, PCR has been one of the most important genetic tools available to those mapping the human genome and for those attempting to detect pathogenic bacteria, viruses, and protozoa. The PCR has made its way from the research lab to forensic and diagnostic laboratories worldwide. There have been considerable efforts to validate and standardize this tool (17, 38); to become a normal routine task/service performed by reference laboratories (3, 48, 76), and clinical diagnostic and food microbiology laboratories (1, 33, 38, 45, 54, 79, 87).

THERMOCYCLER TECHNOLOGY

Since the technique of PCR was developed, there have been many evolutions of the equipment that makes the process possible, based on the concept that strands of DNA denature, or unwind, and anneal, or wind again back into the helical corkscrew, in response to fluctuations in temperature. The first successful PCR reaction took place using water baths at the appropriate temperatures for each step in the procedure, with the technician moving vials by hand from bath to bath at the appointed time, for 30 or more cycles to get adequate amounts of DNA copies that could be detected. Nowadays, thanks to automation, PCR reactions can be set up in thermocyclers that over the course of minutes to a few hours reliably yield high numbers of a specific DNA sequence if present in a sample.

The standard thermocycler uses a large heating block, into which microcentrifuge tubes are placed. This type of thermocycler through its computer controls the heating and coolings of the blocks through the cycles of each reaction. Sometimes an oil or wax overlay is put on the samples within the microcentrifuge tubes to keep the sample from escaping the bottom of the tube during the heating for the denature step of the PCR reaction. This type of machine is not as desirable because of the time it takes to heat and then cool the entire block to the appropriate temperature within each cycle. The time required for the heat block to uniformly reach each temperature coupled with the slow heat transfer rates to the microcentrifuge tubes makes this type of thermocycler virtually inadequate with today's demand for high-speed accurate amplification of PCR products.

The RapidCycler, manufactured by Idaho Technologies, is an example of equipment designed to provide the quick temperature cycling necessary for

PCR reactions. This type of thermocycler uses heat transfer through blasts of high-velocity hot air to accomplish the temperature transactions from the initial heating of the DNA sample through the annealing of the primers and the extension of the new double strand of DNA by the polymerase. There is overall temperature uniformity within the cavity of the reaction vessel and rapid heat exchange within the sample because individual samples are loaded into micro-capillary tubes or thin walled microcentrifuge tubes for the reactions to take place. This also allows for a smaller overall volume in each reaction tube, thus saving valuable amounts of reaction components such as polymerase and primers. After the PCR cycles are complete, the samples are loaded on an agarose gel that contains ethidium bromide, and viewed under a UV light source.

A gradient thermocycler allows the clinician to optimize each of the three temperatures needed for the denaturing, annealing and extension of new DNA products. Optimization might be required if an existing PCR cycle program cannot be located for detection of a particular gene sequence. Optimizing the PCR reactions is critical to the success of the production of amplicons and is not always the easiest thing to do. The melting temperature can be calculated for the primers when they are made, but the denaturing and annealing temperature of the cycle might have to be determined by educated guess with some trial and error, possibly rerunning the same reaction with many different temperatures before the best fit temperature is found. Luckily, most machines have the ability to reach the different temperatures at the same time (thus the gradient), so the reactions are run at the same time with results from each temperature trial collected at the same time. The block used to hold the samples in this type of machine can be programmed to heat over a gradient of about 20°C range with the annealing temperature for the PCR reactions increased by an increment of 1 or 2°C.

Another thermocycler type offers PCR product detection at the same time as each cycle of the PCR reaction progresses. It allows the technician to track

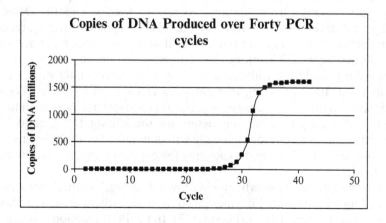

Figure 1.1. Example of real-time PCR output graph showing amount of DNA sequences produced over 40 PCR cycles.

the increase in products during a PCR reaction as displayed on a graph (Fig. 1.1). Higuchi and colleagues introduced this feature, dubbed "Real-Time" PCR, and described how the number of cycles necessary to produce a detectable fluorescence was "directly related to the starting number of DNA copies" in a sample (32). There are a number of companies that offer this technology, which combines the rapid cycle polymerase chain reaction with fluorescent detection of amplified PCR product in the same closed vessel as the reaction mix. The primers are usually labeled with fluorogenic probes, or a DNA-binding dye is included in the PCR reaction, which fluoresces under light emitted at a certain wavelength.

DETECTION

Detection of the PCR product or "amplicon" can be accomplished several ways. Following PCR, the sample is loaded into an agarose gel, and the DNA fragment(s) or amplicon if present in the sample is separated by electrophoresis based on size. Molecular weight, DNA standards are included to estimate size of amplicon(s) present in positive samples and positive control. The agarose gel and electrophoresis buffer contain a dye, ethidium bromide that binds double-stranded DNA and fluoresces upon excitation with UV light. This dye is used to visualize the DNA in an agarose gel. As the primers bind to fixed position within the target sequence, the expected size for our PCR product/amplicon is the distance between the forward and reverse primers. For example, if forward and reverse primers bind target gene X at positions 850 and 1000, respectively, then one expects to observe an amplicon of 150 bp for positive control or any sample that bears organism that contains gene X. The size of the amplicon is extrapolated from the DNA standards included in the gel. The sample MUST produce an amplicon of expected size predicted for the primers used and corrobated by the positive control before it can be considered positive by PCR. There is an inverse linear correlation between the log_{10} size of the DNA fragment (bp) and the distance migrated by the DNA fragment in the agaraose gel. The smaller the DNA fragment the farther it migrates through the agarose gel during electrophoresis. Therefore one can estimate the amplicon's size from DNA standards included with the agarose gel. As most PCRs produce small size amplicons (100–1,000 bp), one must use DNA standards that accommodate this size range and agaraose concentration (1.5%) that resolves small DNA fragments sufficiently to accurately determine the size for DNA band X. The PCR result is recorded photographically with a polaorid or digital camera with the appropriate lens filters and exposures for capturing images illuminated by the UV light.

Detection systems are slowly moving towards nongel methods for detecting and recording PCR results. Enzyme-linked immunosorbent assay (ELISA) has been developed for detecting amplicons (37). In the PCR reaction mix, the standard nucleotides have been substituted for chemically "tagged" nucleotides (e.g., dioxygenin or DIG). During PCR, the "DIG'-labeled nucleotides are

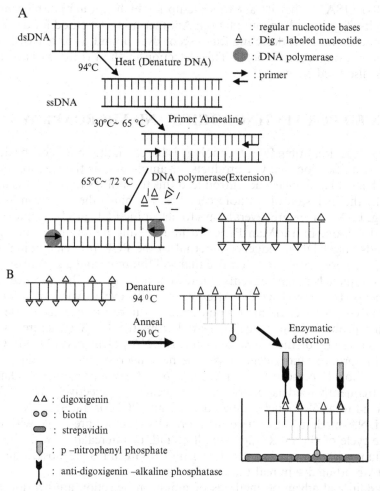

Figure 1.2. PCR-ELISA. Digoxygenin (DIG) labeled, nucleotides are incorporated into amplicon during PCR (A). An internal, 3′ biotinylated oligoprobe anneals to denatured, single-stranded amplicon following PCR. The strepavidin, coating the wells, binds to the biotin moiety of the oligoprobe and thus captures the amplicon. The amplicon is then detected using anti-DIG antibody enzyme conjugate (B). The oligoprobe adds additional specificity to this PCR test.

incorporated into the amplicon as it is synthesized (Fig. 1.2A). Following the last round of PCR, the sample is denatured and allowed to anneal with 5′, biotin-labeled, internal oligonucleotide. This oligoprobe binds to the complementary sequence present within the amplicon. This amplicon-oligoprobe hybrid is captured in strepavidin-coated 96 well microplate through the interaction of the biotin group with the strepavidin (Fig. 1.2B). The bound amplicon is visualized colormeterically using anti-DIG antibody enzyme conjugate, usually either horseradish peroxidase or alkaline phosphatase. The advantage of this "PCR-ELISA" is that it easy to scale-up for high throughput of samples and lends itself quite well to automation. Another nongel method for detecting PCR amplicons involves detecting fluorescent dyes bound to or released from the amplicon using a fluorometer. This detection method is the basis of real-time PCR discussed below.

ADVANCED PCR TECHNOLOGIES AND MICROARRAYS

Real-Time PCR. Real-time PCR technology is based on the ability for the detection and quantification of PCR products, or amplicons, as the reaction cycles progress. Higuchi and colleagues introduced this technology (32) and it is made possible by the inclusion of a fluorescent dye that binds the amplicon as it is made (Fig. 1.3A). There are several ways to detect the PCR products under fluorescent detection. In TaqMan PCR, an intact, "internal" fluorogenic oligoprobe binds target DNA sequence, internal to the PCR primer binding sites. This oligprobe possesses a reporter dye that will fluoresce and a suppressor dye known as a quencher that prevents fluorescent activity via fluorescence resonance energy transfer (FRET). After each PCR cycle, when the double-stranded DNA products are made, a measure of fluorescence is taken after the fluorogenic probe is hydrolytically cleaved from the DNA structure by the exonuclease activity of the *Thermus aquaticus* DNA polymerase (29, 36). Once cleaved, the probe's fluorescent activity is no longer suppressed (Fig. 1.3B). FAM (6-carboxyfluorescein) and TAMRA (6-carboxy-tetramethyl-rhodamin) are most frequently used as reporter and as quencher, respectively. This PCR is often refered to as 5′exonuclease-based, real-time PCR or TaqMan PCR (55). When a DNA-binding dye is used, as more DNA copies are made with each successive cycle of the PCR, they are all bound, or intercalated, with the dye, and the fluorescence increases (Fig. 1.3A). SYBR Green I is the most frequently used DNA-binding dye in real-time PCR.

Two additional advanced methods of amplicon detection are hybridization probes and molecular beacons. The hybridization method uses fluorescence resonance energy transfer from one probe to another after annealing of the primers to the template strand of DNA. One probe has a donor dye at the 3′ end of the oligonucleotide and the other probe has an acceptor dye at the 5′ end. When both probes anneal to the target sequences, they are situated to have the dyes adjacent to one another one base apart. While in that configuration, the energy emitted by the donor dye excites the acceptor dye, which emits fluorescent light at a longer wavelength. The ratio between the two fluorescent emissions increases as the PCR progresses and is proportional to the amount of amplicons produced. Molecular beacons are short segments of single-stranded DNA. They use a hairpin shape to facilitate quenching of the fluorescent signal until the probe anneals to the complementary target DNA sequences, produced from PCR.

Some advantages of real-time technology include high sensitivity with the use of an appropriate probe or DNA-binding dye, ability for detection of relatively small numbers of target DNA copies, and ease of quantification because

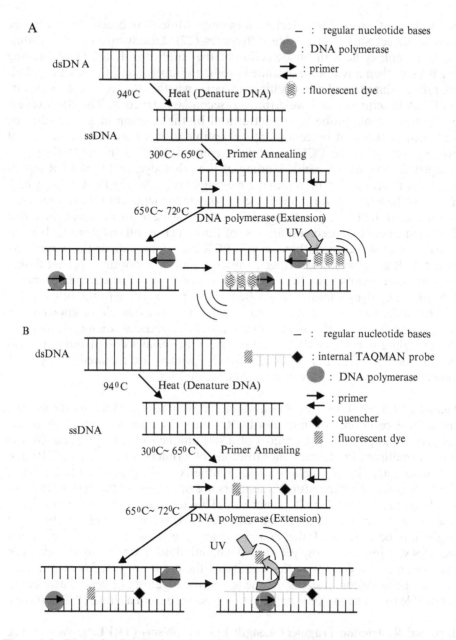

Figure 1.3. "Real-time" PCR detection of amplicons. Using fluorogenic dyes, amplicons can be detected by a fluorimeter as they are synthesized with each PCR cycle. Intially, a fluoresent dye, SYBR Green (A), was used to detect the amplicons. In this PCR, SYBR Green binds the double-stranded, DNA amplicon and fluorescences upon illumination with ultraviolet light (UV). Subsequently, real-time PCR was developed using an internal oligoprobe for detecting the amplicons. In TaqMan PCR (B), the oligoprobe contains a fluorscent marker and a chemical group that quenches fluorescence of the oligoprobe until the dye is liberated by 3′ exonuclease activity of the Taq DNA polymerase. This can only occur if the oligo binds the complementary sequences present in the target gene and amplicon.

of the lack of post-PCR detection measures. Molecular beacons can even be used to detect single nucleotide differences (27). Disadvantages of real-time PCR technology lie with the detection of the amplicons. If the DNA-binding dye is used, then any double-stranded product is labeled and fluoresces, including primer–dimers, and nonspecific amplicons, whether they are close to the target DNA in sequence or have erroneous secondary structure. The effectiveness of the fluorogenic probe is also influenced by the creation of primer–dimers, and both methods of detection are susceptible to less than optimal design of primers for use in the PCR and primer concentration in the master mix. To compensate for the unspecific binding of the DNA dye, real-time PCR equipment has the capability of running a melting curve after the PCR assay, which increases the temperature of the vessels in tiny increments until the fluorescence is lost due to the DNA denaturing. When the melting temperature of the target DNA sequence is reached, a sharp loss of fluorescence will be recorded. If additional losses are recorded, there may be PCR contamination, or the parameters of the PCR assay were not stringent enough, such as suboptimal primer design or temperature choice for the program. This feature of the equipment makes DNA-binding dyes a feasible and often cheaper alternative to the other methods available. The melting temperature of the amplicons should be known when designing primers for the assay and are usually referenced when programming the annealing step of the PCR reaction. Accurate primer design and optimization of the PCR reaction conditions for the primers are required in any PCR application, but especially with real-time technology.

Multiplex PCR. Multiplex PCR is a way to amplify two or more amplicons in a single PCR reaction. For multiplex PCR, each primer set is designed to its target gene to amplify a PCR product of a size unique to the target gene. To perform a multiplex PCR, the concentrations of primers, Mg^{2+}, free dNTPs and polymerase are altered to allow for the synthesis of the genes of interest, while the PCR reaction temperature parameters are optimized to the best average for amplicon production for all primer sets. This technique saves time and labor since more than one target DNA sequence is detected for each reaction, but might not be optimal if the PCR products are close in size and detection requires viewing an agarose gel stained with ethidium bromide. In a single PCR reaction, one can determine the identity of the organism (28, 39) or genotype (21), as the amplicon(s) size is unique to specific organism or gene. Therefore, it is possible to detect multiple pathogens in a sample from a single PCR test (65).

Terminal Restriction Fragment Length Polymorphisms (TRFLP). We can use PCR to characterize microbial communities and identify member species using a single PCR primer set. This PCR targets the 16S rDNA, a gene that is universally conserved among all bacterial species and amplifies a single ~1,500 bp amplicon. We can resolve diversity of 16S rDNA amplicons generated from this PCR using restriction enzyme(s) that recognize restriction sites within genus or species specific sites within this gene and produce DNA fragments, whose size corresponds to a specific genus or species (46). This PCR involves using universal 16s rDNA primers in which one of the forward or reverse primers is fluorescently tagged (Fig. 1.4). Following PCR, the amplicons are digested with a restriction enzyme and subsequently loaded onto capillary bed of an automated DNA sequencer. This method has been refined and applied to automated DNA sequencers to resolve minor (x bp) differences between DNA fragments, monitoring and measuring fluorescence associated with the various sized DNA fragments as they elute from the sequencers' capillary bed (Fig. 1.4). Fluorescently labeled molecular weight standards are included to calibrate column in order to demarcate and identify the molecular weight for each DNA fragment separated by on the sequencer's capillary column. Each peak corresponds to specific genus/species present within the sample. The identity is determined from comparisons to an established database of restriction fragments predicted from 16S rDNA sequences (47). This database can be generated in house, from cloning and sequencing your 16S rDNA library or comparing it against an ever-expanding Web-based 16S rDNA database (Michigan State University Center for Microbial Ecology; *http://35.8.164.52/html/TAP-trflp.html*; 47). The latter has a tool for analyzing your TRFLP profile against this database, for various restriction enzymes. Depending on the restriction enzyme used, one may not be able to resolve various species or genera with a

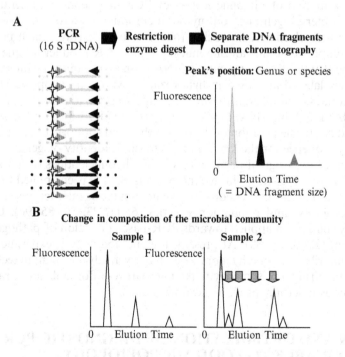

Figure 1.4. Characterizing microbial communities and identification of pathogens in foods from terminal restriction fragment length polymorphisms (TRFLP) of total microbial community 16S rDNA. (A) Concept behind TRFLP. (B) Interpretation of TRFLP.

single restriction enzyme. This is because they produce the same size DNA fragment with restriction enzyme X. It may take a number of different TRFLP profiles of the same community, generated with different restriction enzymes, before genera and/or species differences can be resolved (47). This method is currently used in assessing stability and structure of microbial consortiums, and it has been recently applied to analyzing changes in the community structure of gastrointestinal microflora in response to diet or probiotics (34, 42). TRFLP can also identify signature peaks for microbial pathogens (14, 60), where differences in 16S rDNA can be discerned between them and closely related commensal organisms, exceptions *E. coli* vs. *Salmonella* (61). Theoretically, TFLP and other molecular ecology tools (e.g., DGGE) will prove useful towards analyses of microbial communities present in foods, gastrointestinal tracts of food animals, probiotics and starter cultures and determine the impact certain food processes have on their composition, with regards to the food's safety for consumers.

Microarrays. Macroarrays, microarrays, high-density oligonucleotide arrays, and microelectronic arrays are all part of a new technology that allows one to screen for gene(s), sequence(s) or specific mRNA among myriad of possible sequences or genes in a single test (22). DNA hybridization arrays are based on specific positioning of a myriad of oligonucleotides or PCR amplicons, representative of a complete bacterial genome, on nylon membrane (macroarray), glass slide (microarray), or electronic microchip (microelectronic array). Each position on this solid support contains an oligonucleotide or PCR product unique to a particular gene. Total mRNA or genomic DNA from an organism is fluorescently or radioactively labeled and used in hybridization with solid support. The bound oligonucleotides or amplicons on the solid support serve to capture labeled probe in the RNA: DNA or DNA: DNA hybridization (Fig. 1.5). The labeled nucleic acid hybridizes to the position or "spot" on the solid support that contains complementary sequence for the labeled probe to bind. Identity of gene or sequence relates back to the original positioning of the oligonucleotides or amplicons on the solid support (Fig. 1.5). This technology has already been applied towards the study of bacterial gene expression (30, 71), host-microbe interactions (15, 73, 84), bacterial evolution and population genetics (6, 11, 23, 70, 85, 96). Currently, microarrays have been applied towards PCR-based detection of pathogens in the environment (2, 43, 65, 88). At present, this methodology is experimental, performed primarily by research laboratories. However, advancement in technologies and manufacturing will someday make microarrays affordable and practical for use in diagnostic setting, as PCR has now become.

DESIGN AND OPTIMIZATION OF DIAGNOSTIC PCR AS APPLICABLE TO FOOD MICROBIOLOGY

To perform PCR in any microbiological application, the DNA sequences of an infectious agent must be known, and the target sequences must be unique to the organism(s) to be detected. For example, if a food sample is suspected to be

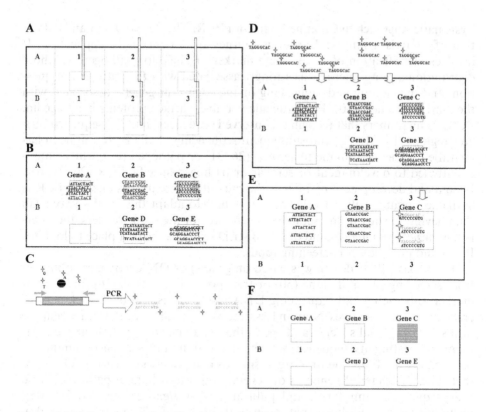

Figure 1.5. Microarrays. Specific oligonucleotides or PCR amplicons are spotted onto defined region on glass side or nylon membrane (A, B). The positioning of this capture probe on this solid matrix defines gene or signature sequence for organism X. If any of the genes present on slide or membrane are present, then it will be amplified during PCR and labeled with fluorescent nucleotide (C) and subsequently bound to the complementary sequence present on the solid support (D, E). Position of fluorescent signal (F) identifies gene or organism present in the sample.

contaminated with bacteria "X", such as *E. coli* O157, then a PCR can be used to determine the presence of the bacteria if there is a gene that only that bacteria possesses, such as an identifier gene "X1", or in the case of *E. coli*, the O157 antigen biosynthesis gene (50). If the gene was found in more than one bacteria type, say gene "XY4," additional PCRs would have to be performed to separate bacteria that harbor that gene (bacteria X and bacteria Y) by looking for a unique identifier gene of the target bacteria X, but at additional work, cost and time for the clinician. The case of identifying *E. coli* O157:H7 might require a multiplex PCR approach because of the closely similar genes of the different antigen subunit serotypes (21, 26). There are many genes that are shared within the same genus and species of bacteria, such as the genes shared among pathogenic *E. coli* strains. Instead of differentiating between bacteria X and Y, the researcher is met with finding a uniqueness of bacteria X1 versus bacteria X2.

Systematic Approach to Creating Your Own PCR. The development and valida-tion of PCR is a long and arduous journey from concept to application. It involves identification of a candidate marker or allele for pathogen X, whose distribution among microbes is strongly associated with the pathogen in ques-tion, and the cloning and sequencing of the cognizant gene(s) associated with the marker or allele (50). For antigenic variable, surface proteins like flagellin, PCR, using primers that recognize conserved sequences flanking sequence vari-able regions (19, 81, 83), and subsequent sequencing of the PCR amplicon has identified sequences unique to serovar (26, 31) or pathogen (63), which subse-quently led to development of serovar or pathogen-specific PCR (26, 31, 63). Design and development of PCR is the pursuit of researchers and if a PCR is available, commercially or otherwise, it is best to adapt this PCR to your lab than having to start from "scratch". Therefore, for most our readers, the inter-net, *www.ncbi.nlm.nih.gov* and the PUBMED search is the best place to look for PCRs and protocols for screening foods.

In the past, PCR design was based on gene(s) or DNA sequences obtained from screening plasmid clone (50) or transposon libraries (5, 16) for relevant marker, subcloning and sequencing DNA inserts. This approach took consid-erable time and resources. Now, in less time, we can sequence the entire genome of a single bacterial species, and spend the remainder of our time at the com-puter annotating its sequence, searching for signature sequences unique to pathogen X. In 1995, the first organism was completely sequenced (20). Since that time, 91 bacterial genomes of several, important human pathogens have been sequenced, annotated, and published (*www.tigr.org*; accessed 2/16/05), including several foodborne pathogens (10, 12, 35, 52, 59, 66, 69, 74, 82, 92). From comparisons of these bacterial genomes, especially between closely related commensals and pathogens, several regions within the chromosome have been identified that appear to be unique to organism X that is tied to its virulence (7, 62, 93), or metabolism (69). With the growing number of bacter-ial genomes present in public accessible DNA databases, identification and design of PCR for organism can be done *in silico*, on your desktop computer. A priori, of course is that organism X's genome has been sequenced and accessible to the user. With advances in PCR and *in silico* analyses of bacter-ial genomes, we can amplify, clone and sequence large regions of the bacterial chromosome to quickly identify target DNA sequences for PCR primer design (89, 90)

Access DNA Databases to Retrieve Sequences or Search for DNA Matches. For the researcher, the most important resource, second only to the library and PUBMED, is the DNA database, GenBank at the National Center for Biotechnology Information, National Institutes of Health, Bethesda, Maryland. This database can be accessed via the internet at the following Website: *www.ncbi.nlm.nih.gov*, go to the ENTREZ selection at the top of the page, and then go to GenBank on the next Web page. One can then search the database of sequences by typing in keywords or combination of words for a specific organism, serovar, or gene(s). Prior to this search, it is important to do your initial research in the library, so that your GenBank search is refined and specific to pull out select sequences from the millions, probably billions, of data base entries present at this Website. The next step is to access a specific GenBank accession, for this exercise we will examine the *Salmonella enterica* Typhimurium LT2 genome at NCBI, GenBank Accession # NC 003917 (Fig. 1.6A, B) and search the annotated genome for the invasion gene *invA* (24, 25) by using the search function in Netscape Navigator for the word "invA" to find the beginning and end of each gene's open reading frame (ORF) (Fig. 1.6C, D). We write down this information and scroll down to the complete sequence to find and copy these sequences (Fig 1.7A). We can paste this sequence for the time being into MSWORD, MSWORDPERFECT, or WORD Notepad and save this file, giving it the organism/gene name. The first three nucleotides should start with ATG, the start codon or rare start codon GTG, and end with TAA, TAG, or TGA, the stop codons. It should be noted, especially with genome sequences, the gene may be in the opposite orientation on the chromosome, requiring inversion of DNA sequence and transcribing the opposite DNA strand to identify start and stop of our ORF. Many DNA software analysis programs can do this for us. We chose the ORF rather than flanking or intergenic regions, because we expect greater selection pressure and less chance for sequence divergence among strains of organism X than these intergenic regions. This is especially important if we are to identify all members of organism X. Now that we have these sequences, we need to determine, *in silico*, whether these sequences are unique to genus *Salmonella* and specifically, the serovar Typhimurium. This can be determined going to BLAST on *www.ncbi.nlm.nih.gov* Website. Click on BLAST and under Nucleotide, click on "nucleotide-nucleotide BLAST (blastn) (Fig. 1.7B)." This will take you to a new site within NCBI that has a box beside "Search". Paste your sequence into this box, and click on the BLAST button (Fig. 1.7B). On the next page, select under the "Format" section, the box titled "or select from" and chose "Bacteria [ORGN]" and click on the FORMAT button (Fig. 1.7C). Allow the BLAST search time to search the database. The time it takes for the search is dependent on gene and the amount of "traffic" at this Website; it is a very popular site with researchers. The results are returned, outlining how many matches there are to your gene sequence. As of 8 February 2005, there were 222 matches with the closest matches, (>90%) to *S. enterica invA* representing various serovars. Other matches are identified, most notably in homologues, genes with similar function, present in *Escherichia coli*. This is expected, as *invA* is part of the type III secretion system present in many human and plant pathogens (40). More importantly, the BLAST results identify for us region of the *invA* sequence to focus on in our primer design. This database search using BLAST is the same approach one would use in analyzing DNA sequence generated from the sequencing of plasmid clones or PCR amplicons. There are however, no guarantees that your gene or sequence will prove useful as a diagnostic marker for organism X, based solely on this database search. You find only what is available on the database, at the time of your search. It is therefore important experimentally to determine the distribution of your candidate gene or allele among

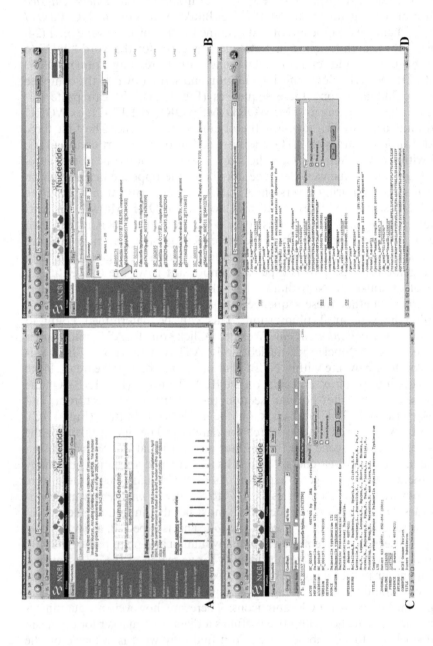

Figure 1.6. National Center for Biotechnology, GenBank DNA database. (A) Entrez. (B) Accessing *Salmonella enterica* Typhimurium LT2 genome GenBank Accession # NC 003917 at NCBI. (C) Search GenBank #NC003917 for "*invA*". (D) Finding the "*invA*" coding sequence (CDS).

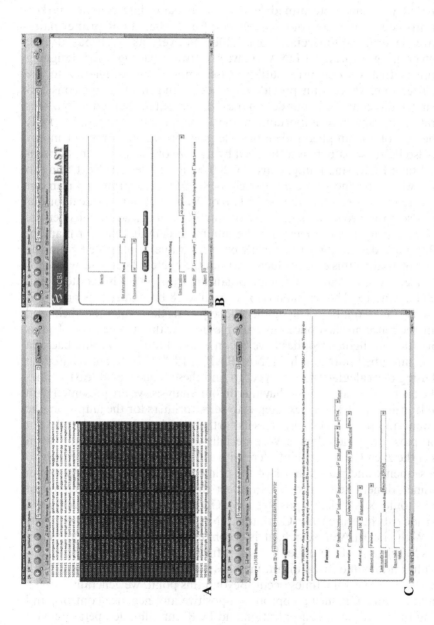

Figure 1.7. National Center for Biotechnology, BLAST Search. (A) Coping "*invA*" coding sequence from GenBank # NC 003917. (B) BLASTn sequence homology search engine. (C) Formatting BLASTn sequence homology search engine.

a sampling of strains, serovars, and closely and distantly related microbes. Despite the presence of *invA* homolog in other genera and species, sequences are divergent enough for this to be a useful genetic marker for detecting *Salmonella* (68, 72).

Now that you have determined *in silico* your candidate genetic marker, you can proceed to analyze your sequence(s) for the best PCR primer pairs. There are several commercially available, as well as Web-based (13; http://dbb.nhri.org.tw/primer/) DNA analysis software packages for designing PCR primers that vary in price, utility, ease, options, or familiarity to the authors. Therefore, we will only provide the reader with general design considerations. First, let us consider in our design the size we want for our amplicon. This consideration is especially important in the development of multiplex PCR where the size of the amplicon identifies the gene or organism present in our sample. Also PCRs sensitivity is influenced by the size of the amplicon. For sensitive, real-time PCR, small amplicons, 75–200 bp are preferred. Next thing to consider is where to concentrate our search for specific PCR primers. From our BLAST search, it appears that the 1st 750 bp of *Salmonella invA* is ideal for our analysis. Also to improve the specificity of our PCR, we need to consider the length of each primer (94). Generally, the minimum default value for many of the PCR primer design algorithms is 18 bp. This value is generated from the probability of finding this exact sequence within the bacterial genome, where for this example; we are dealing with an organism with 50% GC content and 4,000,000 bp genome. The probability of a specific 18 bp sequence is present is $(1/4)^{18} \times 4,000,000 = 6 \times 10^{-5}$. The smaller the sequence, the greater the likelihood of finding sequence not just once but multiple times within the genome. That is why short 10-mer oligonucleotides have become useful tools for typing bacteria by random amplified polymorphic DNA (RAPD) PCR (51), because based on size and using our calculations we expect to find these sequences at least 4 times within the bacterial genome. Now, having run our analysis, we are presented with all possible primer pairs. Our next step is to select primers for the amplicon size that we want and screen these primer sets further, to identify those that do not form "hairpins" or primer–dimers. We especially want to avoid primers that form hairpins at the 3′ end as this will interfere with the primers annealing precisely to its target sequence and participation of the primer in the DNA extension step in PCR. Primer–dimers and hairpins can affect the specificity and sensitivity of PCR and should be avoided if possible (78). Once the appropriate primer set(s) has been identified, search the GenBank DNA database for match with our primers. With the BLAST search, it is recommended with searches of short sequences to select Bacteria under "or select from" option. This is to limit confusion with random and insignificant matches with the larger animal and plant genomes (10^9 bp) that sometimes occur. Beyond this point, we generally empirically optimize our PCR, using appropriate positive and negative controls, and identifying the magnesium concentration and PCR annealing temperature with the sensitivity and specificity that is best for detecting organism X. We then verify the specificity of our PCR by comparing same strains, serovars, or species against different strains, serovars, and closely and distantly related microorganisms to see if same size PCR amplicon is produced only for those groups of bacteria to which the PCR was intended to identify. Ideally, once our PCR has been optimized, PCR amplicon, of the expected size is only observed among select bacteria that possess the target gene and nothing for all other microorganisms that do not possess this gene. It is at this point too that we verify that our amplicon, with size expected based on the primers designed, is the target gene to which our primers were intended to amplify. We accomplish this by sequencing the PCR amplicon and match resulting sequence against GenBank DNA database using the BLAST algorithm. Our amplicon's sequence should match the *invA* sequences present on the DNA database.

REFERENCES

1. Abdulmawjood, A., M. Bulte, S. Roth, H. Schonenbrucher, N. Cook, M. D'Agostino, M. Burkhard, K. Jordan, S. Pelkonen, and J. Hoorfar. 2004. Toward an international standard for PCR-based detection of foodborne *Escherichia coli* O157: Validation of the PCR-based method in a multicenter interlaboratory trial. J. AOAC Int. **87**:856–860.
2. Al-Khaldi, S.F., K.M. Myers, A. Rasooly, and V. Chizhikov. 2004. Genotyping of *Clostridium perfringens* toxins using multiple oligonucleotide microarray hybridization. Mol. Cell Probes **18**:359–367.
3. Anderson, A.D., V.D. Garrett, J. Sobel, S.S. Monroe, R.L. Fankhauser, K.J. Schwab, J.S. Bresee, P.S. Mead, C. Higgins, J. Campana, R.I. Glass, and Outbreak Investigation Team. 2001. Multistate outbreak of Norwalk-like virus gastroenteritis associated with a common caterer. Am. J. Epidemiol. **154**:1013–1019.
4. Benjamin, M.A., J. Lu, G. Donnelly, P. Dureja, and D.M. McKay. 1998. Changes in murine jejunal morphology evoked by the bacterial superantigen *Staphylococcus aureus* enterotoxin B are mediated by CD4+ T cells. Infect. Immun. **66**:2193–2199.
5. Bilge, S.S., J.C. Vary Jr., S.F. Dowell, and P.I. Tarr. 1996. Role of *Escherichia coli* O157:H7 O side chain in adherence and analysis of an *rfb* locus. Infect. Immun. **64**:4795–4801.
6. Borucki, M.K., S.H. Kim, D.R. Call, S.C. Smole, and F. Pagotto. 2005. Selective discrimination of *Listeria monocytogenes* epidemic strains by a mixed-genome DNA microarray compared to discrimination by pulsed-field gel electrophoresis, ribotyping, and multilocus sequence typing. J. Clin. Microbiol. **42**:5270–5276.
7. Brussow, H., C. Canchaya, and W.D. Hardt. 2004. Phages and the evolution of bacterial pathogens: From genomic rearrangements to lysogenic conversion. Microbiol. Mol. Biol. Rev. **68**:560–602.
8. Buzby, J.C. and T. Roberts. 1997. Economic costs and trade impacts of microbial foodborne illness. World Health. Stat. Q. **50**:57–66.
9. Cha, J.H., M.Y. Chang, J.A. Richardson, and L. Eidels. 2003. Transgenic mice expressing the diphtheria receptor are sensitive to the toxin. Mol. Microbiol. **49**:235–240.
10. Chain, P.S., E. Carniel, F.W. Larimer, J. Lamerdin, P.O. Stoutland, W.M. Regala, A.M. Georgescu, L.M. Vergez, M.L Land, V.L. Motin, R.R. Brubaker, J. Fowler, J. Hinnebusch, M. Marceau, C. Medigue, M. Simonet, V. Chenal-Francisque, B. Souza, D. Dacheux, J.M. Elliott, A. Derbise, L.J. Hauser, and E. Garcia. 2004. Insights into the evolution of *Yersinia pestis* through whole-genome comparison with *Yersinia pseudotuberculosis*. Proc. Natl. Acad. Sci. USA **101**:13826–13831.

11. Chan, K., S. Baker, C.C. Kim, C.S. Detweiler, G. Dougan, and S. Falkow. 2003. Genomic comparison of *Salmonella enterica* serovars and *Salmonella bongori* by use of an *S. enterica* serovar Typhimurium DNA microarray. J. Bacteriol. **185**:553–563.

12. Chen, C., K.-M. Wu, Y.-C. Chang, C.-H. Chang, H.-C. Tsai, T.-L. Liao, Y.-M. Liu, H.-J. Chen, A.B.-T. Shen, J. – C. Li, T.-L. Su, C.-P. Shao, C.-T. Lee, L.-I. Hor, and S.-F. Tsai. 2003. Comparative genome analysis of *Vibrio vulnificus*, a marine pathogen. Gen. Res. **13**:2577–2587.

13. Chen, S.H., C.Y. Lin, C.S. Cho, C.Z. Lo, and C.A. Hsiung. 2003. Primer design assistant (PDA): A Web-based primer design tool. Nucl. Acids Res. **31**:3751–3754.

14. Christensen, J.E., J.A. Stencil, and K.D. Reed. 2003. Rapid identification of bacteria from positive blood cultures by terminal restriction fragment length polymorphism profile analysis of the 16S rRNA gene. J. Clin. Microbiol. **41**:3790–3800.

15. Dahan, S., S. Knutton, R.K. Shaw, V.F. Crepin, G. Dougan, and G. Frankel. 2004. Transcriptome of enterohemorrhagic *Escherichia coli* O157 adhering to eukaryotic plasma membranes. Infect. Immun. **72**:5452–5459.

16. Desmarchelier, P.M., S.S. Bilge, N. Fegan, L. Mills, J.C. Vary, and P.I. Tarr. 1998. A PCR specific for *Escherichia coli* O157 based on the *rfb* locus encoding O157 lipopolysaccharide. J. Clin. Microbiol. **36**:1801–1804.

17. Dimech, W., D.S. Bowden, B. Brestovac, K. Byron, G. James, D. Jardine, T. Sloots, and E.M. Dax. 2004. Validation of assembled nucleic acid-based tests in diagnostic microbiology laboratories. Pathol. **36**:45–50.

18. Donohue-Rolfe, A., I. Kondova, S. Oswald, D. Hutto, and S. Tzipori. 2000. *Escherichia coli* O157:H7 strains that express Shiga toxin (Stx) 2 alone are more neurotropic for gnotobiotic piglets than are isotypes producing only Stx1 or both Stx1 and Stx 2. J. Infect. Dis. **181**:1825–1829.

19. Fischer, S.H., and I. Nachamkin. 1991. Common and variable domains of the flagellin gene, *flaA*, in *Campylobacter jejuni*. Mol. Microbiol. **5**:1151–1158.

20. Fraser, C.M., J.D. Gocayne, O. White, M.D. Adams, R.A. Clayton, R.D. Fleischmann, C.J. Bult, A.R. Kerlavage, G. Sutton, J.M. Kelley, et al. 1995. The minimal gene complement of *Mycoplasma genitalium*. Science **270**:397–403.

21. Fratamico, P.M., S.K. Sackitey, M. Wiedmann, and M.Y. Deng. 1995. Detection of *Escherichia coli* O157:H7 by multiplex PCR. J. Clin. Microbiol. **33**:2188–2191.

22. Freeman, W.M., D.J. Robertson, and K.E. Vrana. 2000. Fundamentals of DNA hybridization arrays for gene expression analysis. Biotechniques **29**:1042–1055.

23. Fukiya, S., H. Mizoguchi, T. Tobe, and H. Mori. 2004. Extensive genomic diversity in pathogenic *Escherichia coli* and *Shigella* strains revealed by comparative genomic hybridization microarray. J. Bacteriol. **186**:3911–3921.

24. Galan, J.E. and R. Curtiss III. 1989. Cloning and molecular characterization of genes whose products allow *Salmonella typhimurium* to penetrate tissue culture cells. Proc. Natl. Acad. Sci. USA **86**:6383–6387.

25. Galan, J.E. and R. Curtiss III. 1991. Distribution of the *invA*, *-B*, *-C*, and *-D* genes of *Salmonella typhimurium* among other *Salmonella* serovars: *invA* mutants of *Salmonella Typhi* are deficient for entry into mammalian cells. Infect. Immun. **59**:2901–2908.

26. Gannon, V.P., S. D'Souza, T. Graham, R.K. King, K. Rahn, and S. Read. 1997. Use of the flagellar H7 gene as a target in multiplex PCR assays and improved specificity in identification of enterohemorrhagic *Escherichia coli* strains. J. Clin. Microbiol. **35**:656–662.

27. Giesendorf, B.A.J., J.A.M. Vet, S. Tyagi, E.J.M.G. Mensink, F.J.M. Trijbels, and H.J. Blom. 1998. Molecular beacons: A new approach for semiautomated mutation analysis. Clin. Chem. **44**:482–486.

28. Harmon, K.M., G.M. Ransom, and I.V. Wesley. 1997. Differentiation of *Campylobacter jejuni* and *Campylobacter coli* by polymerase chain reaction. Mol. Cell. Probes **11**:195–200.

29. Heid, C.A., J. Stevens, K.J. Livak, and P.M. Williams. 1996. Real time quantitative PCR. Genome Res. **6**:986–994.

30. Herold, S., J. Siebert, A. Huber, and H. Schmidt. 2005. Global expression of prophage genes in *Escherichia coli* O157:H7 strain EDL933 in response to norfloxacin. Antimicrob. Agents Chemother. **49**:931–944.

31. Herrera-Leon, S., J.R. McQuiston, M.A. Usera, P.I. Fields, J. Garaizar, and M.A. Echeita. 2004. Multiplex PCR for distinguishing the most common phase-1 flagellar antigens of *Salmonella spp. J.* Clin. Microbiol. **42**:2581–2586.

32. Higuchi, R., C. Fockler, G. Dollinger, and R. Watson. 1993. Kinetic PCR analysis: Real-time monitoring of DNA amplification reactions. Biotechnol. **11**:1026–1030.

33. Hirsch, H.H. and W. Bossart. 1999. Two-centre study comparing DNA preparation and PCR amplification protocols for herpes simplex virus detection in cerebrospinal fluids of patients with suspected herpes simplex encephalitis. J. Med. Virol. **57**:31–35.

34. Hogberg, A., J.E. Lindberg, T. Leser, and P. Wallgren. 2004. Influence of cereal non-starch polysaccharides on ileo-caecal and rectal microbial populations in growing pigs. Acta. Vet. Scand. **45**:87–98.

35. Holden, M.T., E.J. Feil, J.A. Lindsay, S.J. Peacock, N.P. Day, M.C. Enright, T.J. Foster, C.E. Moore, L. Hurst, R. Atkin, A. Barron, N. Bason, S.D. Bentley, C. Chillingworth, T. Chillingworth, C. Churcher, L. Clark, C. Corton, A. Cronin, J. Doggett, L. Dowd, T. Feltwell, Z. Hance, B. Harris, H. Hauser, S. Holroyd, K. Jagels, K. D. James, N. Lennard, A. Line, R. Mayes, S. Moule, K. Mungall, D. Ormond, M. A. Quail, E. Rabbinowitsch, K. Rutherford, M. Sanders, S. Sharp, M. Simmonds, K. Stevens, S. Whitehead, B.G. Barrell, B.G. Spratt, and J. Parkhill. 2004. Complete genomes of two clinical *Staphylococcus aureus* strains: Evidence for the rapid evolution of virulence and drug resistance. Proc. Natl. Acad. Sci. USA **101**:9786–9791.

36. Holland, P.M., R.D. Abramson, R. Watson, D.H. Gelfand. 1991. Detection of specific polymerase chain reaction product by utilizing the 5′–3′ exonuclease activity of *Thermus aquaticus* DNA polymerase. Proc. Natl. Acad. Sci. USA **88**:7276–7280.

37. Hong, Y., M.E. Berrang, T. Liu, C.L. Hofacre, S. Sanchez, L. Wang, and J.J. Maurer. 2003. Rapid detection of *Campylobacter coli, C. jejuni*, and *Salmonella enterica* on poultry carcasses by using PCR-enzyme-linked immunosorbent assay. Appl. Environ. Microbiol. **69**:3492–3499.

38. Hoorfar, J., P. Wolffs, and P. Radstrom. 2004. Diagnostic PCR: Validation and sample preparation are two sides of the same coin. APMIS **112**:808–814.

39. Houf, K., A. Tutenel, L. De Zutter, J. Van Hoof, and P. Vandamme. 2000. Development of a multiplex PCR assay for the simultaneous detection and identification of *Arcobacter butzleri, Arcobacter cryaerophilus,* and *Arcobacter skirrowii.* FEMS Microbiol. Lett. **193**:89–94.

40. Hueck, C.J. 1998. Type III protein secretion systems in bacterial pathogens of animal and plants. Microbiol. Mol. Biol. Rev. **62**:379–433.

41. Johnson, W.M., S.D. Tyler, E.P. Ewan, F.E. Ashton, D.R. Pollard, and K.R. Rozee. 1991. Detection of genes for enterotoxins, exfoliative toxins, and toxic shock syndrome toxin 1 in *Staphylococcus aureus* by the polymerase chain reaction. J. Clin. Microbiol. **29**:426–430.

42. Kaplan, C.W., J.C. Astaire, M.E. Sanders, B.S. Reddy, and C.L. Kitts. 2001. 16S ribosomal DNA terminal restriction fragment patter analysis of bacterial communities

in feces of rats fed *Lactobacillus acidophilus* NCFM. Appl. Environ. Microbiol. **67**:1935–1939.

43. Keramas, G., D.D. Bang, M. Lund, M. Madsen, H. Bunkenborg, P. Telleman, and C. B. V. Christensen. 2004. Use of culture, PCR analysis, and DNA microarrays for detection of *Campylobacter jejuni* and *Campylobacter coli* from chicken feces. J. Clin. Microbiol. **42**:3985–3991.

44. Koopmans, M. and E. Duizer. 2004. Foodborne viruses: An emerging problem. Int. J. Food Microbiol. **90**:23–41.

45. Mackay, I.M. 2004. Real-time PCR in the microbiology laboratory. Clin. Microbiol. Infect. **10**:190–212.

46. Marsh, T.L. 1999. Terminal restriction fragment length polymorphism (T-RFLP): An emerging method for characterizing diversity among homologous populations of amplification products. Curr. Opin. Microbiol. **2**:323–327.

47. Marsh, T.L., P. Saxman, J. Cole, and J. Tiedje. 2000. Terminal restriction fragment length polymorphism analysis program, a Web-based research tool for microbial community analysis. Appl. Environ. Microbiol. **66**:3616–3620.

48. Marston, C.K., F. Jamieson, F. Cahoon, G. Lesiak, A. Golaz, M. Reeves, and T. Popovic. 2001. Persistence of a distinct *Corynebacterium diphtheriae* clonal group within two communities in the United States and Canada where diphtheria is endemic. J. Clin. Microbiol. **39**:1586–1590.

49. Maurer, J.J. and M.D. Lee. 2005. *Salmonella. In* Mansel Griffiths (ed.), Understanding Pathogen Behavior, Woodhead Publishing Limited, Cambridge, United Kingdom (In Press).

50. Maurer, J.J., D. Schmidt, P. Petrosko, S. Sanchez, L. Bolton, and M.D. Lee. 1999. Development of primers to O-antigen biosynthesis genes for specific detection of *Escherichia coli* O157 by PCR. Appl. Env. Microbiol. **65**:2954–2960.

51. Mazurier, S., A. van de Giessen, K. Heuvelman, and K. Wernars. 1992. RAPD analysis of *Campylobacter* isolates: DNA fingerprinting without the need to purify DNA. Lett. Appl. Microbiol. **14**:260–262.

52. McClelland, M., K.E. Sanderson, J. Spieth, S.W. Clifton, P. Latreille, L. Courtney, S. Porwollik, J. Ali, M. Dante, F. Du, S. Hou, D. Layman, S. Leonard, C. Nguyen, K. Scott, A. Holmes, N. Grewal, E. Mulvaney, E. Ryan, H. Sun, L. Florea, W. Miller, T. Stoneking, M. Nhan, R. Waterston, and R. K. Wilson. 2001. Complete genome sequence of *Salmonella enterica* serovar Typhimurium LT2. Nature **413**:852–856.

53. Mead, P.S., L. Slutsker, V. Dietz, L.F. McCaig, J.S. Bresee, C. Shapiro, P.M. Griffin, and R.V. Tauxe. 1999. Food-related illness and death in the United States. Emerg. Infect. Dis. **5**:607–625.

54. Moscoso, H., S.G. Thayer, C.L. Hofacre, and S.H. Kleven. 2004. Inactivation, storage, and PCR detection of Mycoplasma on FTA filter paper. Avian Dis. **48**:841–850.

55. Mullah, B., K. Livak, A. Andrus, and P. Kenney. 1998. Efficient synthesis of double dye-labeled oligodeoxyribonucleotide probes and their application in a real time PCR assay. Nucleic Acids Res. **26**:1026–1031.

56. Mullis, K.B., F.A. Faloona, S.J. Scharf, R.K. Saiki, G.T. Horn, and H.A. Erlich. 1986. Specific enzymatic amplification of DNA in vitro: The polymerase chain reaction. Cold Spring Harbor Symp. Quant. Biol. **51**:263–273.

57. Mullis, K.B. 1994. *In* K.B. Mullis, F. Ferré, and R.A. Gibbs (ed.), The Polymerase Chain Reaction. Birkhäuser Boston, Cambridge, MA.

58. Mullis, K.B. 1998. Dancing Naked in the Mind Field. Pantheon Books, New York.

59. Nelson, K.E., D.E Fouts., E.F.Mongodin, J. Ravel, R.T. DeBoy, J.F. Kolonay, D.A. Rasko, S.V. Angiuoli, S.R. Gill, I.T. Paulsen, J. Peterson, O. White, W.C. Nelson, W.

Nierman, M.J. Beanan, L.M. Brinkac, S.C. Daugherty, R.J. Dodson, A.S. Durkin, R. Madupu, D.H. Haft, J. Selengut, S. Van Aken, H. Khouri, N. Fedorova, H. Forberger, B. Tran, S. Kathariou, L.D. Wonderling, G.A. Uhlich, D.O. Bayles, J.B. Luchansky, and C.M. Fraser. 2004. Whole genome comparisons of serotype 4b and 1/2a strains of the foodborne pathogen *Listeria monocytogenes* reveal new insights into the core genome components of this species. Nucl. Acids Res. **32**:2386–2395.

60. Nilsson, W.B. and M.S. Strom. 2002. Detection and identification of bacterial pathogens of fish in kidney tissue using terminal restriction fragment length polymorphism (T-RFLP) analysis of 16S rRNA genes. Dis. Aquat. Organ. **48**:175–185.

61. Nordentoft, S., H. Cristensen, and H.C. Wegener. 1997. Evaluation of a fluorescence-labeled oligonucleotide probe targeting 23S rRNA for in situ detection of *Salmonella* serovars in paraffin-embedded tissue sections and their rapid identification in bacterial smears. J. Clin. Microbiol. **35**:2642–2648.

62. Oelschlaeger, T.A. and J. Hacker. 2004. Impact of pathogenicity islands in bacterial diagnostics. APMIS **112**:930–936.

63. Oyofo, B.A., S.A. Thornton, D.H. Burr, T.J. Trust, O.R. Pavlovskis, and P. Guerry. 1992. Specific detection of *Campylobacter jejuni* and *Campylobacter coli* by using polymerase chain reaction. J. Clin. Microbiol. **30**:2613–2619.

64. Panet, A. and H.G. Khorana. 1974. Studies on polynucleotides. The linkage of deoxyribonucleic templates to cellulose and its use in their replication. J. Biol. Chem. **249**:5213–5221.

65. Panicker, G., D.R. Call, M.J. Krug, and A.K. Bej. 2004. Detection of pathogenic *Vibrio spp.* in shellfish by using multiplex PCR and DNA microarrays. Appl. Environ. Microbiol. **70**:7436–7444.

66. Parkhill, J., B.W. Wren, K. Mungall, J.M. Ketley, C. Churcher, D. Basham, T. Chillingworth, R.M. Davies, T. Feltwell, S. Holroyd, K. Jagels, A.V. Karlyshev, S. Moule, M.J. Pallen, C.W. Penn, M.A Quail, M.A. Rajandream, K.M. Rutherford, A.H. van Vliet, S. Whitehead, and B.G. Barrell. 2000. The genome sequence of the foodborne pathogen *Campylobacter jejuni* reveals hypervariable sequences. Nature **403**:665–668.

67. Paton, A.W. and J.C. Paton. 1998. Detection and characterization of shiga toxigenic *Escherichia coli* by using multiplex PCR assays for stx_1, stx_2, *eae*A, enterohemorrhagic *E. coli hly*A, rfb_{O111}, and rfb_{O157}. J. Clin. Microbiol. **36**:598–602.

68. Perelle, S., F. Dilasser, B. Malorny, J. Grout, J. Hoorfar, and P. Fach. 2004. Comparison of PCR-ELISA and LightCycler real-time PCR assays for detecting *Salmonella spp.* in milk and meat samples. Mol. Cell Probes **18**:409–420.

69. Perna, N.T., G. Plunkett 3rd, V. Burland, B. Mau, J.D. Glasner, D.J. Rose, G.F. Mayhew, P.S. Evans, J. Gregor, H.A. Kirkpatrick, G. Posfai, J. Hackett, S. Klink, A. Boutin, Y. Shao, L. Miller, E.J. Grotbeck, N.W. Davis, A. Lim, E.T. Dimalanta, K.D. Potamousis, J. Apodaca, T.S. Anantharaman, J. Lin, G. Yen, D.C. Schwartz, R.A. Welch, and F.R. Blattner. 2001. Genome sequence of enterohaemorrhagic *Escherichia coli* O157:H7. Nature **409**:529–533.

70. Porwollik, S., E.F. Boyd, C. Choy, P. Cheng, L. Florea, E. Proctor, and M. McClelland. 2004. Characterization of *Salmonella enterica* subspecies I genovars by use of microarrays. J. Bacteriol. **186**:5883–5898.

71. Prouty, A.M., I.E. Brodsky, S. Falkow, and J.S. Gunn. 2004. Bile-salt-mediated induction of antimicrobial and bile resistance in *Salmonella typhimurium*. Microbiol. **150**:775–783.

72. Rahn, K., S.A. Grandis, R.C. Clarke, S.A. McEwen, J.E. Galan, C. Ginocchio, R. Curtiss III, and C.L. Gyles. 1992. Amplification of *invA* gene sequence of *Salmonella*

typhimurium by polymerase chain reaction as a specific method of detection of *Salmonella*. Mol. Cell Probes **6**:271–279.

73. Rappuoli, R. 2000. Pushing the limits of cellular microbiology: Microarrays to study bacteria-host cell intimate contacts. Proc. Natl. Acad. Sci. USA **97**:13467–13469.

74. Rasko, D.A., J. Ravel, O.A. Okstad, E. Helgason, R.Z. Cer, L. Jiang, K.A. Shores, D.E. Fouts, N.J. Tourasse, S.V. Angiuoli, J. Kolonay, W.C. Nelson, A.B. Kolsto, C.M. Fraser, and T.D. Read. 2004. The genome sequence of *Bacillus cereus* ATCC 10987 reveals metabolic adaptations and a large plasmid related to *Bacillus anthracis* pXO1. Nucl. Acids Res. **32**:977–988.

75. Reeves, M.W., R.J. Arko, F.W. Chandler, and N.B. Bridges. 1986. Affinity purification of staphylococcal toxic shock syndrome toxin 1 and its pathologic effects in rabbits. Infect. Immun. **51**:431–439.

76. Robertson, B.H., F. Averhoff, T.L. Cromeans, X. Han, B. Khoprasert, O.V. Nainan, J. Rosenberg, L. Paikoff, E. DeBess, C.N. Shapiro, and H.S. Margolis. 2000. Genetic relatedness of hepatitis A virus isolated during a community-wide outbreak. J. Med. Virol. **62**:144–150.

77. Rotbart, H.A., M.J. Levin, L.P. Villarreal, S.M. Tracy, B.L. Semler, and E. Wimmer. 1985. Factors affecting the detection of enteroviruses in cerebrospinal fluid with coxsackievirus B3 and poliovirus 1 cDNA probes. J. Clin. Microbiol. **22**:220–224.

78. Rychlik, W. 1995. Priming efficiency in PCR. Biotechn. **18**:84–86, 88–90

79. Saldanha, J., W. Gerlich, N. Lelie, P. Dawson, K. Heermann, A. Heath, and WHO Collaborative Study Group. 2001. An international collaborative study to establish a World Health Organization international standard for hepatitis B virus DNA nucleic acid amplification techniques. Vox Sang. **80**:63–71.

80. Savitt, E.D., M.N. Strzempko, K.K. Vaccaro, W.J. Peros, and C.K. French. Comparison of cultural methods and DNA probe analyses for the detection of *Actinobacillus actinomycetemcomitans*, *Bacteroides gingivalis*, and *Bacteroides intermedius* in subgingival plaque samples. J. Periodontol. **59**:431–438.

81. Schoenhals, G. and C. Whitfield. 1993. Comparative analysis of flagellin sequences from *Escherichia coli* strains possessing serologically distinct flagellar filaments with a shared complex surface pattern. J. Bacteriol. **175**:5395–5402.

82. Shimizu, T., K. Ohtani, H. Hirakawa, K. Ohshima, A. Yamashita, T. Shiba, N. Ogasawara, M. Hattori, S. Kuhara, and H. Hayashi. 2002. Complete genome sequence of *Clostridium perfringens*, an anaerobic flesh-eater. Proc. Natl. Acad. Sci. USA **99**:996–1001.

83. Smith, N.H. and R.K. Selander. 1990. Sequence invariance of the antigenic-coding central region of the phase 1 flagellar filament gene (*fliC*) among strains of *Salmonella typhimurium*. J. Bacteriol. **172**:603–609.

84. Stintzi, A., D. Marlow, K. Palyada, H. Naikare, R. Panciera, L. Whitworth, and C. Clarke. 2005. Use of genome-wide expression profiling and mutagenesis to study the intestinal lifestyle of *Campylobacter jejuni*. Infect. Immun. **73**:1797–1810.

85. Taboada, E.N., R.R. Acedillo, C.D. Carrillo, W.A. Findlay, D.T. Medeiros, O.L. Mykytczuk, M.J. Roberts, C.A. Valencia, J.M. Farber, and J.H. Nash. 2004. Large-scale comparative genomics meta-analysis of *Campylobacter jejuni* isolates reveals low level of genome plasticity. J. Clin. Microbiol. **42**:4566–4576.

86. Thomas, E.J., R.K. King, J. Burchak, and V.P. Gannon. 1991. Sensitive and specific detection of *Listeria monocytogenes* in milk and ground beef with the polymerase chain reaction. Appl. Environ. Microbiol. **57**:2576–2580.

87. Vinje, J., H. Vennema, L. Maunula, C.H. von Bonsdorff, M. Hoehne, E. Schreier, A. Richards, J. Green, D. Brown, S.S. Beard, S.S. Monroe, E. de Bruin, L. Svensson,

and M.P. Koopmans. 2003. International collaborative study to compare reverse transcriptase PCR assays for detection and genotyping of noroviruses. J. Clin. Microbiol. **41**:1423–1433.

88. Vora, G.J., C.E. Meador, D.A. Stenger, and J.D. Andreadis. 2004. Nucleic acid amplification strategies for DNA microarray-based pathogen detection. **70**:3047–3054.

89. Wang, L., W. Qu, and P.R. Reeves. 2001. Sequence analysis of four *Shigella boydii* O-antigen loci: Implication for *Escherichia coli* and *Shigella* relationships. Infect. Immun. **69**:6923–6930.

90. Wang, L., and P.R. Reeves. 1998. Organization of *Escherichia coli* O157 O antigen gene cluster and identification of its specific genes. Infect. Immun. **66**:3545–3551.

91. Watson, J.D., and F.H. Crick. 1953. Molecular structure of nucleic acids; a structure for deoxyribose nucleic acid. Nature. **171**:737–738.

92. Wei, J., M.B. Goldberg, V. Burland, M.M. Venkatesan, W. Deng, G. Fournier, G.F. Mayhew, G. Plunkett 3rd, D.J. Rose, A. Darling, B. Mau, N.T. Perna, S.M. Payne, L. J. Runyen-Janecky, S. Zhou, D.C. Schwartz, and F.R. Blattner. 2003. Complete genome sequence and comparative genomics of *Shigella flexneri* serotype 2a strain 2457T. Infect. Immun. **71**:2775–2786.

93. Whittam, T.S., and A.C. Bumbaugh. 2002. Inferences from whole-genome sequences of bacterial pathogens. Curr. Opin. Genet. Dev. **12**:719–725.

94. Wu, D.Y., L. Ugozzoli, B.K. Pal, J. Qian, and R.B. Wallace. 1991. The effect of temperature and oligonucleotide primer length on the specificity and efficiency of amplification by the polymerase chain reaction. DNA Cell. Biol. **10**:233–238

95. Yoo, H.S., S.U. Lee, K.Y. Park, and Y.H. Park. 1997. Molecular typing and epidemiological survey of prevalence of *Clostridium perfringens* types by multiplex PCR. J. Clin. Microbiol. **35**:228–232.

96. Zhang, C., M. Zhang, J. Ju, J. Nietfeldt, J. Wise, P.M. Terry, M. Olson, S.D. Kachman, M. Wiedmann, M. Samadpour, and A. K. Benson. 2003. Genome diversification in phylogenetic lineages I and II of *Listeria monocytogenes*: Identification of segments unique to lineage II populations. J. Bacteriol. **185**:5573–5584.

97. Zhang, X., A.D. McDaniel, L.E. Wolf, G.T. Keusch, M.K. Waldor, and D.W. Acheson. 2000. Quinolone antibiotics induce Shiga toxin-encoding bacteriophages, toxin production, and death in mice. J. Infect. Dis. **181**:664–670.

CHAPTER 2

The Mythology of PCR:
A Warning to the Wise

John J. Maurer[1,2]*, Ph.D.

*Poultry Diagnostic & Research Center[1], College of Veterinary Medicine,
The University of Georgia, Athens, GA 30602
Center for Food Safety[2], College of Agriculture and Environmental Sciences,
The University of Georgia, Griffin, GA 30223*

Introduction
Interpretation
  Conventional PCR
  Real Time PCR
Validation
Problems and Their Solutions
  False-Positives and Dead vs. Live Bacterial Cell Debate
  PCR Inhibitors, Limits of Detection and False-Negatives
Conclusions
References

INTRODUCTION

Most diagnostic PCR tests are a qualitative yes or no, presence or absence of pathogen X. We know what it means if our sample is positive by PCR, reporting back presumptive positive for organism X and a negative PCR result was the end-point for that sample. Were these assumptions correct? The decisions we make based on these PCR results require that we know how to interpret these results and like any other diagnostic test, know its limitations with regards to sensitivity and specificity. Even if your laboratory is only interested in adapting existing PCR methods for identification of pathogens in foods, it is important that you know what the results mean, and know how to recognize and troubleshoot problems as they occur. You can safe guard or at least be prepared to recognize these problems, as they appear, by implementing standard operating procedures and including controls recommended by authors in the chapters discussed in this book. In this section, I will specifically delve into interpretation and understanding of PCR results as well as discuss the limitations, problems, and erroneous assumptions associated with PCR and other PCR based technologies (e.g., real-time PCR).

*Corresponding author. Phone: (706) 542-5071; FAX: (706) 542-5630; e-mail: jmaurer@vet.uga.edu.

INTERPRETATION

Conventional PCR. A sample is positive, by PCR, if an amplicon is produced with the size expected for the primers used. What if the sample yields an amplicon larger or smaller than the size expected for our PCR primers? Is this sample considered positive by PCR? NO!!! This result is referred to in PCR parlance as a nonspecific amplicon, it is ignored, AND *if we do not observe an amplicon with a size expected for primers used, the sample is considered PCR negative*. It is therefore a requirement to always include DNA molecular weight standards, in the appropriate size range for accurately assessing the amplicon's size, and the percentage of agaraose and electrophoresis time needed to adequately separate the molecular weight standards. One needs to also consider other parameters (electrophoresis buffer, buffer strength, voltage, etc.) that affect uniformity of DNA separation across the entire width and length of the agarose gel. For wide gels with many wells or lanes (>10), one may consider placement of the DNA standards in the middle and the outermost wells. With appropriate gel documentation software, the user can, using these well placed molecular weight standards, correct for electrophoresis migration anomalies that produce "smiles" at high voltages. Avoiding electrophoresis at high voltages or circulate/cool the buffer during electrophoresis can prevent this electrophoretic anomaly. With every PCR run, ALWAYS include a positive control so that you can match your sample with the control, and allow adequate separation of your DNA standards, samples, and control so that you do not erroneously report a sample with a nonspecific amplicon as positive. If molecular biology is new to your laboratory, it is advisable to purchase a general molecular biology manual, that details the specifics of gel electrophoresis, includes theory and helps trouble shoot problems commonly associated with the molecular technique (1, 62).

For the experienced molecular biologist, this is rather obvious, but for others, especially the novice, it is easy to be lulled into believing the presence of any PCR product, regardless of size, on the gel means the sample must be positive for organism X. Most genes targeted by PCR have been selected based on their conservation and uniformity within a species, subspecies, serovar or pathotype. These genes are uniform in size. There are, however, exceptions, genes or DNA segments containing repetitive elements or extragenic sequences, the number, size, or presence of which varies within the bacterial population (10, 16, 22, 38, 57, 70). PCRs have been developed to exploit these genetically variable regions for the purpose of genus/species identification (10, 16, 24, 35, 57) and strain typing (25, 56, 57). Here the different size amplicon identifies the genus or species and/or distinguishes strain types. However, a requirement for using any of these PCRs is first the isolation of the organism. For PCR screens of foods, it is advisable to avoid those PCRs that produce, as designed, these variable size amplicons. Unless, an internal probe is included in the PCR screens, for specificity, the technician may confuse a true, nonspecific amplicon in a sample as a positive and erroneously report the sample as such.

Real-Time PCR. Results generated by real-time PCR are generally more straightforward to interpret for a simple question like: is the organism present

in our sample? Rather than visualize the amplicon following PCR, we monitor
the increase in fluorescence over time as newly synthesized, amplicon binds to
SYBR Green® or the chemically quenched, fluorescent dye is liberated as the
amplicon displaces an internally bound, dye-labeled probe. Fig. 2.1 illustrates
kinetics of real-time PCR. Note the points on the x-axis, "threshold cycle" (C_T)
where the log-linear phase of fluorescence begins for the different target DNA
concentrations (43). There is a linear correlation between C_T and DNA concentration, making the PCR quantitative. A sample is considered positive provided it falls within the range of C_T values, demarcated by the PCR's limit of
detection, and the background fluorescence associated with the negative or no
DNA controls. While real-time PCR surpasses conventional PCR in speed and
sensitivity, nonspecific amplicons can result in our erroneously reporting a positive result. SYBR Green® binds to double stranded DNA, regardless of
whether it is the expected amplicon, nonspecific amplicon, or primer-dimers.
Gradient thermocyclers have become a useful tool in rapidly identifying annealing temperature best for PCR amplification of the target gene while avoiding
primer-dimers. We can distinguish nonspecific amplicon(s) from a true positive
based on their distinctive DNA melting curves (Fig. 2.2) (59).

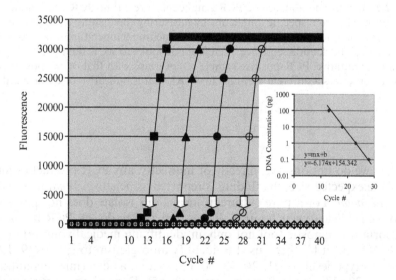

Figure 2.1. Detection of foodborne pathogen X in foods by real time PCR. As amplicon is synthesized, the thermocycler continuously measures fluorescences with each
cycle. The PCR product fluoresces due to binding of fluorescent dye, SYBER Green
to the double stranded DNA, amplicon as it is formed. When the PCR amplicons are
first detected during real time, is a function of the target DNA concentration: ■ (100
pg), ▲ (10 pg), ● (1 pg), (0.1 pg), ◇ (0.01 pg), and + NO DNA control. Arrows identify the "threshold cycle," C_T on the x-axis, # PCR cycles where the log-linear phase of
fluorescence begins. The cycle numbers the target DNA concentration was plotted relative to C_T and as shown in the inset, there is a negative, linear correlation between
DNA concentration and C_T.

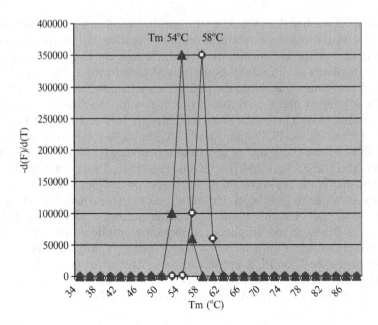

Figure 2.2. Identifying nonspecific PCR amplicons in real time PCR. We can distinguish nonspecific from specific amplicons by measuring the melting temperature (Tm) for each amplicon following real-time PCR. The melting temperature is a reflection of the amplicons's nucleotide sequence, therefore one looks to see if the DNA melting curve for the putative, PCR-positive sample (□) overlaps with that of the positive control (♦) or produces a different melting curve (▲), that is indicative of a nonspecific amplicon.

When we do not observe, directly or indirectly, any PCR product or amplicon of the expected size, the finding is reported as negative. *What does a negative result mean?* For a pure culture, it means our isolate does not possess the gene or gene allele to which our PCR was designed to detect. PCR has become an important diagnostic tool not only in identifying medically important genera (40, 58), but it has been used to identify an organism to species (9, 19, 23, 40), or serotype level (6, 21, 26, 42, 50, 66) as well as determine its antibiotic resistance (20, 27) or virulence potential (2, 55). Depending on the organism and gene(s) or gene alleles associated with resistance to drug X, PCR negative result may indicate: (1) the organism is susceptible to the antibiotic in question (e.g., mycobacterium and isonazid resistance; 27); or (2) PCR negative only means the organism does not possess this gene (e.g.,enterococci and streptogramin resistance; 69) and susceptibility cannot be inferred. Gene screens to assess, genotypically, drug resistance is challenging due to multiple genes and gene alleles associated with resistance to certain antibiotics (8, 15, 64). With regards to detection of multi-drug resistant (MDR) pathogens, while it is tempting to select antibiotic resistance genes associated with the MDR as probes in PCR screens (31), mobile genetic elements have disseminated these drug resistance genes to innocuous commensals also contaminating foods (32, 63, 74), providing potential for false positives. Gene screens for these MDR loci should therefore be limited to the cultured pathogen. For detection of pathogens in foods, it is imperative that we select a target gene or sequence that is unique to pathogen X, uniform in its distribution within this bacterial population and genetically stable.

If this target gene is strongly associated with genus, species or serotype, a negative PCR translates to this isolate is NOT the species, strain, or serotype identified by this PCR. However, if we apply this very same PCR to screen for the presence of the organism that the PCR was designed to identify, does a negative result mean it is NOT present? We now are confronted with several questions relating to our PCR test's sensitivity and specificity (28, 39), important in assessing, validating and finally standardizing our PCR for screening pathogens in foods (30).

VALIDATION

In optimizing any PCR, we strive to design, identify and develop the primer set(s) for discerning the one genus, species, or strain from multitude of microbial species while being able to detect the fewest cells possible. This is the molecular biology definition of specificity and sensitivity, respectively. To determine specificity, we test our PCR against, many different bacterial strains, closely or distantly related species and/or genera. A PCR specific for *Salmonella*, for example, will produce positive results, amplicon of the expected size, for ALL *Salmonella* species, strains, and serovars but will prove negative for all other bacterial species, especially closely related species (28, 58). If we continue using *Salmonella* as our example, sensitivity is measured by lowest *Salmonella* cell density detectable by our PCR (28, 39). In its infancy, PCR's specificity and sensitivity were determined using pure cultures and at best a food product was spiked with the offending organism and PCR was performed to detect the organism in the processed sample. Only recently have investigators vigorously put PCR through its paces in the real world to validate its utility for rapid detection of pathogens in foods (30).

Validation of any diagnostic PCR involves comparison against another test, considered the "gold standard" for detection. For food microbiologists, the "gold standard" is the bacteriological approach of culture, isolation, and the biochemical or serological confirmatory tests. From this comparison, we determine statistical specificity (false-positives) and sensitivity (false-negatives) of our PCR test (28, 39). A false-positive is when the sample is PCR positive but culture negative, while a false-negative is vice-versa: PCR negative, culture positive. *What is responsible for reporting false-positives and false-negatives and what can we do to minimize this in our food microbiology lab?*

PROBLEMS AND THEIR SOLUTIONS

False positives can be attributed to several things, most you cannot control, but at least one you can: PCR, template, or sample contamination. As discussed in Chapter 4, "Making PCR a Normal Routine of the Food Microbiology Lab," preventative measures and standard operating procedures are essential to avoid these contamination issues. These measures include physically separating DNA and PCR preparation areas from each other as well as from the area where gel-electrophoresis is performed; use of barrier tips, disposable gloves; and cleaning the PCR preparation area with bleach and/or overnight, ultraviolet illumination. As mentioned earlier, PCR amplifies target gene 10^9-fold, producing more than enough molecules per pg-fg of template to serve as template in the next PCR reaction. Following a PCR run and upon opening the tube, we create an aerosol of amplicons that can quickly contaminate our hands, pipettes, and the immediate bench area. Something as simple as disposing of our gloves following the loading of our PCR sample in the agarose gel and before we set up our next PCR reaction, can avoid future PCR "carry-over" contamination. PCR contamination results in considerable down time for the diagnostic laboratory due to the time it takes to identify the source of contamination, and subsequent decontamination of the affected area or disposal of the contaminated reagent(s). Alternatively, some labs substitute thymidine with uracil in the PCR reaction mix and subsequent pretreatment of all PCR reaction mixes with uracil DNA glycosylase prior to running these reactions in the thermocycler (41). The principle behind this method is that during PCR, amplicons incorporate uracil; the amplicon is now susceptible to hydrolysis by uracil DNA glycosylase, and eliminated prior to each subsequent PCR run. Therefore, erroneous reporting of false-positives due to PCR contamination is eliminated.

As synthesis of the amplicons is identified in "real-time" with newer, PCR fluorescence-based detection technologies, tubes never have to be opened following the initial PCR reaction set-up. With conventional PCR, we can identify PCR contamination when negative or NO DNA controls turn positive. For an experienced lab, something is amiss, when the number of PCR positives greatly exceeds frequency the lab normally encounters for this PCR test or incidence reported in the literature AND subsequent culture results do not correlate with the PCR (i.e., increase in false positives). This can be observed with real-time PCR as lower C_T than encountered for past PCR-positive samples, indicative of high-cell density or template/target concentration, and fails to yield the organism upon culture. As PCR can be extremely sensitive, great care must be taken in sample preparation to avoid cross-contamination. Inclusion of a negative control, sample prep with every PCR run will be useful in identifying cross-contamination, as a positive PCR for this negative control would definitely be indicative of template/sample contamination. Anytime when its evident there is PCR contamination, discard the results for that PCR run, discontinue any future PCRs, and identify and correct the problem immediately before rerunning PCR on any past or future samples.

Nonspecific PCR amplicons can also result in erroneously reporting a sample positive for pathogen X. This can be especially problematic for real-time

PCR using SYBR Green® to detect PCR products and multiplex PCR: the multiple gene screens, single PCR test where the size of the amplicon identifies: genus, species, serotype, or strain. For real-time PCR, we can identify this problem by measuring the amplicon's Tm from the DNA melting peak as stated earlier or run sample PCR on gel side-by-side with positive control to see any differences between the two in their size. Tweaking the PCR conditions to improve its stringency can sometimes eliminate these nonspecific amplicons. This can be done by empirically identify annealing temperature or $MgCl_2$ concentration, that eliminates signal for our "false-positive" sample while not affecting the positive control. To increase the stringency of the PCR, one increases the annealing temperatures and/or lower the $MgCl_2$ concentrations in the reaction mix.

Another way we can improve the specificity of our PCR and reporting false positives, is to use PCR that incorporates internal: nested PCR primers (39); DNA: DNA hybridization or capture probes (28, 45); molecular beacon (37); or TaqMan probe (29). These PCRs have improved specificity because the internal capture or detection probes can distinguish between the real amplicon and nonspecific amplicons, by binding to the complementary sequence within the target amplicon during PCR or at DNA: DNA hybridization step. These internal probes also heighten the sensitivity of the PCR at least 100-fold (28, 45).

False-Positives and Dead vs. Live Bacterial Cell Debate. Even when PCR is running optimally, there may not always be 100% agreement between PCR and culture. The reasons for false-positives are not completely understood. Several explanations have been offered and include: (1) the bacteria are in a viable, nonculturable state (61); (2) injury of the bacterial cells during food processing (52); or (3) the bacteria are dead (48). One can obtain variable culture results alone depending on: (1) whether to include a step(s) that allows for the recovery of injured cells (13, 33); (2) the type of enrichment broth (18) and culture conditions (12) used, or (3) the use of a delayed, secondary enrichment (72, 73) and may explain the disconnect sometimes observed between PCR and culture results.

Depending on where samples are taken within the continuum of food processing steps, especially at Critical Control Point (CCP) designed to reduce or eliminate the pathogen, (e.g., heating), PCR may not be able to distinguish live, dead or damaged cells. In fact, we routinely boil bacteria or wash cells in ethanol to prepare template for PCR, conditions that readily and rapidly kill bacteria. Therefore, one may consider where and when to use PCR in assessing product contamination with pathogen X. For a process that readily ruptures or dissolves the bacterial cell, pre-DNAse treatment step can remove residual DNA carried over from dead, lysed cells (51). However, a significant proportion of heat-treated cells remain intact, dead, and suitable as template for PCR (51). We still need to know whether CCP was effective at eliminating the pathogen or reducing it to an acceptably safe level. PCR still affords us the opportunity to identify the few cells still viable following CCP step, (e.g., pasteurization), by using RNA as the template. Unlike DNA, RNA has short-half life in the bacteria cell (34), as genes are turned on and off as the cell grows and responds to its environment.

Upon cell death, these mRNA transcripts are quickly degraded. There has been considerable interest in using RNA as the template for diagnostic PCR to detect the few viable cells remaining in the sample (17, 46, 52, 65, 75). This can be accomplished by converting RNA to its complementary (c) DNA copy with the retroviral reverse transcriptase, at which point the cDNA can serve as template in standard PCR. This procedure is referred to as reverse-transcriptase PCR. The challenge currently is identifying a constitutively, expressed gene that has sequence unique to the organism and has a short, mRNA half-life, especially upon death of the bacterial cell (75). RNA turnover in the bacterial cell is dependent on its intracellular ribonucleases, and like any enzyme once denatured it becomes inactive and the RNA therefore persists, which may explain the long half-life of RNAs following thermal inactivation of the bacterial cell (47). Therefore, there are times when culture continues to be necessary in assessing microbial risk following food processing step at CCP and other instances where PCR trumps culture in the detection of foodborne pathogens (see below).

Finally, we are left to consider *viable but nonculturable* (VBNC) bacteria and PCR. We know bacteria can enter a physiological state where, with the microscope, we know they are present and viable, as determined using viability stains, but we are unable to plate them from sample X. This VBNC state may result from cellular injury (14), adaptation to a harsh, oxygen-poor or nutrient deplete environment (5, 7, 71) or subsequent transformation from planktonic to sessile state in biofilms (11). In foods, the VBNC state may be the consequence of cellular injury/damage and may require a recovery period, in a preenrichment broth, before the cells can be cultivated. Organisms like *Vibrio* and *Campylobacter* can readily enter VBNC state, especially in aquatic environments (54, 60). Although regarded as a foodborne pathogen, *Campylobacter* is also recognized as the cause of several waterborne outbreaks in the United States (4, 36). With *Campylobacter*, the VBNC state may be due to physical or chemical agent that damages the cell, or nutrient depletion or limitation triggers a physiological change to a survival state. When *Campylobacter* enters the VBNC state, its cell morphology changes from helical to coccoidal. Pathogens can revert back from this nongrowing, VBNC state into actively growing; cultivatable state, under the right conditions in vitro (7, 71) and cause disease in its animal host (53). It may be that we are unable to detect it in this state using our current selective, enrichment media because of the antibiotics in the media that interfere with cellular repair and changes to the cell wall necessary to resume growth (67, 68). Where our culture-based approaches currently have failed, PCR offers the opportunity for the pathogen's detection, especially in its VBNC state (3, 49, 52).

PCR Inhibitors, Limits of Detection, and False-Negatives. False-negatives, PCR-negative, culture positive samples are attributed to two major factors: PCR inhibitors or the PCR's limit of detection. PCR inhibitors may be attributed to the food sample itself or the enrichment used to amplify the target organism. We can often remove these inhibitors by using simple DNA affinity, spin columns to produce clean DNA template for PCR, making samples generally recalcitrant to PCR (e.g., soil) pliable for PCR-based screens and analyses.

Chapter 4, "Sample Preparation for PCR" will go into more detail concerning sample preparation and preparation of template that is free of PCR inhibitors. More recently, diagnostic PCRs for screening foods have been adapted to include an "internal control" in the sample screened in order to eliminate possibility of extraneous factors (e.g., PCR inhibitors) from factoring into interpretation of PCR negative results. "Internal amplification control" is the cloned, positive-control amplicon where an internal region has been removed (44). As template in PCR, "internal amplification control" produces a smaller sized-amplicon. The plasmid DNA bearing our "internal amplification control" is included with sample template in PCR. If the sample is negative for organism X, a single amplicon, corresponding in size to that expected for the "internal amplification control." However, for a positive sample, two amplicons are produced; one that corresponds in size to that expected from amplification of the organism X's targeted gene and the other corresponds in size to that expected for the internal control.

For most PCR beginners, false-negatives due to PCR's insensitivity to detect a single-cell per sample appear to be a paradoxical, if not a heretical statement. You have probably read many research papers and believe their claim that their PCR can detect a single cell/ml of a sample. Is this really possible? With PCR, we are generally dealing with reaction mix volumes that range between 10 and 100 µl to which we may add 1 or 10 µl of the sample, once its been processed for PCR. What is the probability that you detect 1cell/ml by PCR, if you were to take 0.001 ml or 1 µl, once from that sample? Knowing Poisson distribution, we know that odds are very small that we can detect it. However, if we took multiple aliquots from this same sample, a most-probable number approach, we would improve our chances of detecting this organism by PCR. The reality is that for most PCRs the limit of detection is 1–1000 cells per 1 µl sample template run, which translates to $1,000-1 \times 10^6$ cells/ml. Therefore, if we relied on PCR alone, and discounting PCR inhibitors, does a PCR negative sample mean the organism is NOT present? Ideally, one wants to use the PCR that is the most sensitive for identifying pathogen X in our food product. How might we improve our chances of detecting our pathogen knowing these limitations and assuming the organism might be present in our specimens at levels <1000 cells/ml? One approach is to concentrate cells into a smaller volume, or include an enrichment step that amplifies what few cells are present to levels above the PCR's threshold for detection (Chapter 3). For the latter, short enrichment period may be sufficient to bring cell density of the pathogen above the detection limits of the PCR. Enrichments have been especially adapted to PCR protocols for foods due to the necessity of processing the large sample volumes associated with screening foods from pathogens.

CONCLUSIONS

One must be aware of the limitation of any diagnostic test, and PCR is no exception. Will PCR soon be the substitute for current culture or immunological tests for foodborne pathogens? Probably not for all pathogens, but it will become standard tool for detecting foodborne protozoans and viruses, pathogens that are currently recalcitrant to culture-based methods of detection. PCR will become an important tool in identification of serotypes and pathotypes. It can be a useful part of any detection scheme, helping with decisions as to which samples and enrichments to focus our efforts towards (39). Of course, acceptance and implementation of PCR in the diagnostic laboratory requires an understanding of its mechanics, meaning of results, the test's limitations, and being able to recognize problems and trouble-shoot them as they arise.

REFERENCES

1. Ausubel, F.M., R. Brent, R.E. Kingston, D.D. Moore, J.G. Seidman, J.A. Smith, and K. Struhl. 1992. Short Protocols in Molecular Biology, 2nd edn. John Wiley & Sons, New York.
2. Bhaduri, S., B. Cottrell, and A.R. Pickard. 1997. Use of a single procedure for selective enrichment, isolation, and identification of plasmid-bearing virulence *Yersinia enterocolitica* of various serotypes from pork. Appl. Environ. Microbiol. **63**:1657–1660.
3. Binsztein, N., M.C. Costagliola, M. Pichel, V. Jurquiza, F.C. Ramirez, R. Akselman, M. Vacchino, A. Huq, and R. Colwell. 2004. Viable but nonculturable *Vibrio cholerae* O1 in the aquatic environment of Argentina. Appl. Environ. Microbiol. **70**:7481–7486.
4. Blackburn, B.G., G.F. Craun, J.S. Yoder, V. Hill, R.L. Calderon, N. Chen, S.H. Lee, D.A. Levy, and M.J. Beach. 2004. Surveillance for waterborne-disease outbreaks associated with drinking water—United States, 2001–2002. MMWR Surveill. Summ. **53**:23–45.
5. Boaretti, M., M. del Mar Lieo, B. Bonato, C. Signoretto, and P. Canepari. 2003. Involvement of *rpoS* in the survival of *Escherichia coli* in the viable but nonculturable state. Environ. Microbiol. **5**:986–996.
6. Borucki, M.K. and D.R. Call. 2003. *Listeria monocytogenes* serotype identification by PCR. J. Clin. Microbiol. **41**:5537–5540.
7. Bovill, R.A. and B.M. Mackey. 1997. Resuscitation of 'nonculturable' cells from aged cultures of *Campylobacter jejuni*. Microbiol. **143**:1575–1581.
8. Bradford, P.A. 2001. Extended-spectrum beta-lactamases in the 21st century: characterization, epidemiology, and detection of this important resistance threat. Clin. Microbiol. Rev. **14**:933–951.
9. Bubert, A., I. Hein, M. Rauch, A. Lehner, B. Yoon, W. Goebel, and M. Wagner. 1999. Detection and differentiation of *Listeria spp.* by a single reaction based on multiplex PCR. Appl. Environ. Microbiol. **65**:4688–4692.
10. Bubert, A., S. Kohler, and W. Goebel. 1992. The homologous and heterologous regions within the *iap* gene allow genus- and species-specific identification of *Listeria spp.* by polymerase chain reaction. Appl. Environ. Microbiol. **58**:2625–2632.
11. Buswell, C.M., Y.M. Herlihy, L.M. Lawrence, J.T. McGuiggan, P.D. Marsh, C.W. Keevil, and S.A. Leach. 1998. Extended survival and persistence of *Campylobacter spp.* in water and aquatic biofilms and their detection by immunofluorescent-antibody and rRNA staining. Appl. Environ. Microbiol. **64**:733–741.
12. Carlson, V.L., G.H. Snoeyenbos, B.A. McKie, and C.F. Smyser. 1967. A comparison of incubation time and temperature for the isolation of *Salmonella*. Avian Dis. **11**:217–225.

13. Chang, V.P., E.W. Mills, and C.N. Cutter. 2003. Comparison of recovery methods for freeze-injured *Listeria monocytogenes*, *Salmonella* Typhimurium, and *Campylobacter coli* in cell suspensions and associated with pork surfaces. J. Food Prot. **66**:798–803.

14. Chaveerach, P., A.A. ter Huurne, L.J. Lipman, and F. van Knapen. 2003. Survival and resuscitation of ten strains of *Campylobacter jejuni* and *Campylobacter coli* under acid conditions. Appl. Environ. Microbiol. **69**:711–714.

15. Chopra, I. and M. Roberts. 2001. Tetracycline antibiotics: Mode of action, applications, molecular biology, and epidemiology of bacterial resistance. Microbiol. Mol. Biol. Rev. **65**:232–260.

16. Christensen, H., K. Jorgensen, and J.E. Olsen. 1999. Differentiation of *Campylobacter coli* and *C. jejuni* by length and DNA sequence of the 16S-23S rRNA internal spacer region. Microbiol. **145**:99–105.

17. Coutard, F., M. Pommepuy, S. Loaec, and D. Hervio-Heath. 2005. mRNA detection by reverse transcription-PCR for monitoring viability and potential virulence in a pathogenic strain of *Vibrio parahaemolyticus* in viable but nonculturable state. J. Appl. Microbiol. **98**:951–961.

18. D'Aoust, J.Y., A.M. Sewell, and D.W. Warburton. 1992. A comparison of standard cultural methods for the detection of foodborne *Salmonella*. Int. J. Food Microbiol. **16**:41–50.

19. Di Pinto, A., G. Ciccarese, G. Tantillo, D. Catalano, and V.T. Forte. 2005. A collagenase-targeted multiplex PCR assay for identification of *Vibrio alginolyticus*, *Vibrio cholerae*, and *Vibrio parahaemolyticus*. J. Food Prot. **68**:150–153.

20. E-Hajj, H.H., S.A. Marras, S. Tyagi, F.R. Kramer, and D. Alland. 2001. Detection of rifampin resistance in *Mycobacterium tuberculosis* in a single tube with molecular beacons. J. Clin. Microbiol. **39**:4131–4137.

21. Fratamico, P.M., S.K. Sackitey, M. Wiedmann, and M.Y. Deng. 1995. Detection of *Escherichia coli* O157:H7 by multiplex PCR. J. Clin. Microbiol. **33**:2188–2191.

22. Frenay, H.M.E., J.P.G. Theelen, L.M. Schouls, C.M.J.E. Vandenbroucke-Grauls, J. Verhoef, W.J. van Leeuwen, and F.R. Mooi. 1994. Discrimination of epidemic and nonepidemic methicillin-resistant *Staphylococcus aureus* strains on the basis of protein A gene polymorphisms. J. Clin. Microbiol. **32**:846–847.

23. Gonzalez, I., K.A. Grant, P.T. Richardson, S.F. Park, and M.D. Collins. 1997. Specific identification of the enteropathogens *Campylobacter jejuni* and *Campylobacter coli* by using a PCR test based on the *ceuE* gene encoding a putative virulence determinant. J. Clin. Microbiol. **35**:759–763.

24. Graham, T., E.J. Golsteyn-Thomas, V.P. Gannon, and J.E. Thomas. 1996. Genus- and species-specific detection of *Listeria monocytogenes* using polymerase chain reaction assays targeting the 16S/23S intergenic spacer region of the rRNA operon. Can. J. Microbiol. **42**:1155–1162.

25. Gurtler, V. 1993. Typing of *Clostridium difficile* strains by PCR-amplification of variable length 16S-23S rDNA spacer regions. J. Gen. Microbiol. **139**:3089–3087.

26. Herrera-Leon, S., J.R. McQuiston, M.A. Usera, P.I. Fields, J. Garaizar, and M.A. Echeita. 2004. Multiplex PCR for distinguishing the most common phase-1 flagellar antigens of *Salmonella spp*. J. Clin. Microbiol. **42**:2581–2586.

27. Herrera-Leon, L., T. Molina, P. Saiz, J.A. Saez-Nieto, and M.S. Jimenez. 2005. New multiplex PCR for rapid detection of isoniazid-resistant *Mycobacterium tuberculosis* clinical isolates. Antimicrob. Agents Chemother. **49**:144–147.

28. Hong, Y., M. Berrang, T. Liu, C. Hofacre, S. Sanchez, L. Wang, and J.J. Maurer. 2003. Rapid detection of *Campylobacter coli*, *C. jejuni* and *Salmonella enterica* on

poultry carcasses using PCR-enzyme-linked immunosorbent assay. Appl. Environ. Microbiol. **69**:3492–3499.

29. Hoorfar, J., P. Ahrens, and P. Radstrom. 2000. Automated 5' nuclease PCR assay for identification of *Salmonella enterica*. J. Clin. Microbiol. **38**:3429–3435.

30. Hoorfar, J., P. Wolffs, and P. Radstrom. 2004. Diagnostic PCR: Validation and sample preparation are two sides of the same coin. APMIS **112**:808–814.

31. Khan, A.A., M.S. Nawaz, S.A. Khan, and C.E. Cerniglia. 2000. Detection of multidrug-resistant *Salmonella typhimurium* DT104 by multiplex polymerase chain reaction. FEMS Microbiol. Lett. **182**:355–360.

32. Keyes, K., C. Hudson, J.J. Maurer, S.G. Thayer, D.G. White, and M.D. Lee. 2000. Detection of florfenicol resistance genes in *Escherichia coli* isolated from sick chickens. Antimicrob. Agents Chemother. **44**:421–424.

33. Knabel, S.J., H.W. Walker, P.A. Hartman, and A.F. Mendonca. 1990. Effects of growth temperature and strictly anaerobic recovery on the survival of *Listeria monocytogenes* during pasteurization. Appl. Environ. Microbiol. **56**:370–376.

34. Kushner, S.R. 1996. mRNA decay, pp. 849–860. *In* F.C. Neidhardt (ed.), *Escherichia coli* and *Salmonella*: Cellular and Molecular Biology. ASM Press, Washington, DC.

35. Kuwahara, T., I. Norimatsu, H. Nakayama, S. Akimoto, K. Kataoka, H. Arimochi, and Y. Ohnishi. 2001. Genetic variation in 16S-23S rDNA internal transcribed spacer regions and the possible use of this genetic variation for molecular diagnosis of *Bacteroides* species. Microb. Immunol. **45**:191–199.

36. Lee, S.H., D.A. Levy, G.F. Craun, M.J. Beach, and R.L. Calderon. 2002. Surveillance for waterborne-disease outbreaks—United States, 1999–2000. Morb. Mort. Week. Rep. **51**(SS-8):1–66.

37. Liming, S.H. and A.A. Bhagwat. 2004. Application of a molecular beacon-real-time PCR technology to detect *Salmonella* species contaminating fruits and vegetables. Int. J. Food Microbiol. **95**:177–187.

38. Liu, T., M. Garcia, S. Levisohn, D. Yogev, and S.H. Kleven. 2001. Molecular variability of the adhesion-encoding gene *pvpA* among *Mycoplasma gallisepticum* strains and its application in diagnosis. J. Clin. Microbiol. **39**:1882–18888.

39. Liu, T., K. Liljebjelke, E. Bartlett, C.L. Hofacre, S. Sanchez, and J.J. Maurer. 2002. Application of nested PCR to detection of *Salmonella* in poultry environments. J. Food Prot. **65**:1227–1232.

40. Logan, J.M., K.J. Edwards, N.A. Saunders, and J. Stanley. 2001. Rapid identification of *Campylobacter spp.* by melting peak analysis of bioprobes in real-time PCR. J. Clin. Microbiol. **39**:2227–2232.

41. Longo, M.C., M.S. Berninger, and J.L. Hartley. 1990. Use of uracil DNA glycosylase to control carry-over contamination in polymerase chain reactions. Gene **93**:125–128.

42. Luk, J.M., U. Kongmuang, P.R. Reeves, and A.A. Lindberg. 1993. Selective amplification of abequose and paratose synthase genes (*rfb*) by polymerase chain reaction for identification of *Salmonella* major serogroups (A, B, C2, and D). J. Clin. Microbiol. **31**:2118–2123.

43. Mackay, I.M., K.E. Arden, and A. Nitsche. 2002. Real-time PCR in virology. Nucleic Acids Res. **30**:1292–1305.

44. Malorny, B., J. Hoorfar, C. Bunge, and R. Helmuth. 2003. Multicenter validation of the analytical accuracy of *Salmonella* PCR: Towards an international standard. Appl. Environ. Microbiol. **69**:290–296.

45. Maurer, J.J., D. Schmidt, P. Petrosko, S. Sanchez, L. Bolton, and M.D. Lee. 1999. Development of primers to O-antigen biosynthesis genes for specific detection of *Escherichia coli* O157 by PCR. Appl. Env. Microbiol. **65**:2954–2960.

46. McIngvale, S.C., D. Elhanafi, and M.A. Drake. 2002. Optimization of reverse transcriptase PCR to detect viable shiga-toxin producing *Escherichia coli*. Appl. Environ. Microbiol. **68**:799–806.
47. McKillip, J.L., L.A. Jaykus, and M. Drake. 1998. rRNA stability in heat-killed and UV-irradiated enterotoxigenic *Staphylococcus aureus* and *Escherichia coli* O157:H7. Appl. Environ. Microbiol. **64**:4264–4268.
48. McKillip, J.L., L.A. Jaykus, and M. Drake. 1999. Nucleic acid persistence in heat-killed *Escherichia coli* O157:H7 from contaminated skim milk. J. Food Prot. **62**:839–844.
49. Moore, J., P. Caldwell, and B. Millar. 2001. Molecular detection of *Campylobacter spp.* in drinking, recreational and environmental water supplies. Int. J. Hyg. Environ. Health **204**:185–189.
50. Nachamkin, I., H. Ung, and C.M. Patton. 1996. Analysis of HL and O serotypes of *Campylobacter* strains by flagellin gene typing system. J. Clin. Microbiol. **34**:277–281.
51. Nogva, H.K., A. Bergh, A. Holck, and K. Rudi. 2000. Application of the 5'-nuclease PCR assay in evaluation and development of methods for quantitative detection of *Campylobacter jejuni*. Appl. Environ. Microbiol. **66**:4029–4936.
52. Novak, J.S., and V.K. Juneja. 2001. Detection of heat injury in *Listeria monocytogenes* Scott A. J. Food Prot. **64**:1739–1743.
53. Oliver, J.D., and R. Bockian. 1995. In vivo resuscitation, and virulence towards mice, of viable but nonculturable cells of *Vibrio vulnificus*. Appl. Environ. Microbiol. **61**:2620–2623.
54. Oliver, J.D., F. Hite, D. McDougald, N.L. Andon, and L.M. Simpson. 1995. Entry into, and resuscitation from, the viable but nonculturable state by *Vibrio vulnificus* in an estuarine environment. Appl. Environ. Microbiol. **61**:2624–2630.
55. Pass, M A., R. Odedra, and R.M. Batt. 2000. Multiplex PCRs for identification of *Escherichia coli* virulence genes. J. Clin. Microbiol. **38**:2001–2004.
56. Payne, R.E., M.D. Lee, D.W. Dreesen, and H.M. Barnhart. 1999. Molecular epidemiology of *Campylobacter jejuni* in broiler flocks using randomly amplified polymorphic DNA-PCR and 23S rRNA-PCR and role of litter in its transmission. Appl. Environ. Microbiol. **65**:260–263.
57. Pourcel, C., Y. Vidgop, F. Ramisse, G. Vergnaud, and C. Tram. 2003. Characterization of a tandem repeat polymorphism in *Legionella pneumophila* and its use for genotyping. J. Clin. Microbiol. **41**:1819–1826.
58. Rahn, K., S.A. Grandis, R.C. Clarke, S.A. McEwen, J.E. Galan, C. Ginocchio, R. Curtiss III, and C.L. Gyles. 1992. Amplification of *invA* gene sequence of *Salmonella typhimurium* by polymerase chain reaction as a specific method of detection of *Salmonella*. Mol. Cell Probes **6**:271–279.
59. Ririe, K.M., R.P. Rasmussen, and C.T. Wittwer. 1997. Product differentiation by analysis of DNA melting curves during the polymerase chain reaction. Anal. Biochem. **245**:154–160.
60. Rollins, D.M. and R.R. Colwell. 1986. Viable but nonculturable stage of *Campylobacter jejuni* and its role in survival in the natural aquatic environment. Appl. Environ. Microbiol. **52**:531–538.
61. Sails, A.D., F.J. Bolton, A.J. Fox, D.R. Wareing, and D.L. Greenway. 2002. Detection of *Campylobacter jejuni* and *Campylobacter coli* in environmental waters by PCR enzyme-linked immunosorbent assay. Appl. Environ. Microbiol. **68**:1319–1324.
62. Sambrook, J., E.F. Fritsch, and T. Maniatis. 1989. Molecular Cloning: A Laboratory Manual, 2nd edn. Cold Spring Harbor Laboratory Press, Cold Spring Harbor, NY.
63. Sanchez, S., M.A. McCrackin Stevenson, C.R. Hudson, M. Maier, T. Buffington, Q. Dam, and J.J. Maurer. 2002. Characterization of multi-drug resistant *Escherichia coli* associated with nosocomial infections in dogs. J. Clin. Microbiol. **40**:3586–3595.

64. Shaw, K.J., P.N. Rather, R.S. Hare, and G.H. Miller. 1993. Molecular genetics of aminoglycoside resistance genes and familial relationships of the aminoglycoside-modifying enzymes. Microbiol. Rev. **57**:138–163.
65. Sheridan, G.E.C., E.A. Szabo, and B.M. Mackey. 1999. Effect of post-treatment holding conditions on detection of *tufA* mRNA in ethanol-treated *Escherichia coli*: Implications for RT-PCR-based indirect viability tests. Lett. Appl. Microbiol **29**:375–379.
66. Shi, F., Y.Y. Chen, T.M. Wassenaar, W.H. Woods, P.J. Coloe, and B.N. Frye. 2002. Development and application of a new scheme for typing *Campylobacter jejuni* and *Campylobacter coli* by PCR-based restriction fragment length polymorphism analysis. J. Clin. Microbiol. **40**:1791–1797.
67. Signoretto, C., M.M. Lleo, and P. Canepari. 2002. Modification of the peptidoglycan of *Escherichia coli* in the viable but nonculturable state. Curr. Microbiol. **44**:125–131.
68. Signoretto, C., M.M. Lleo, M.C. Tafi, and P. Canepari. 2000. Cell wall chemical composition of *Enterococcus faecalis* in the viable but nonculturable state. Appl. Environ. Microbiol. **66**:1953–1959.
69. Simjee, S., D.G. White, J. Meng, D.D. Wagner, S. Qaiyumi, S. Zhao, J.R. Hayes, and P.F. McDermott. 2002. Prevalence of streptogramin resistance genes among *Enterococcus* isolates recovered from retail meats in the Greater Washington DC area. J. Antimicrob. Chemother. **50**:877–882.
70. Spinaci, C., G. Magi, C. Zampaloni, L.A. Vitali, C. Paoletti, M.R. Catania, M. Prenna, L. Ferrante, S. Ripa, P.E. Varaldo, and B. Facinelli. Genetic diversity of cell-invasive erythromycin-resistant and –susceptible group A streptococci determined by analysis of *RD2* region of the *prtF1* gene. J. Clin. Microbiol. **42**:639–644.
71. Wai, S.N., Y. Mizunoe, A. Takade, and S. Yoshida. 2000. A comparison of solid and liquid media for resuscitation of starvation- and low-temperature-induced nonculturable cells of *Aeromonas hydrophila*. Arch. Microbiol. **173**:307–310.
72. Waltman, W.D., A. Horne, C. Pirkle, and T. Dickson. 1991. Use of delayed secondary enrichment for the isolation of *Salmonella* in poultry and poultry environments. Avian Dis. **35**:88–92.
73. Waltman, W.D., A.M. Horne, C. Pirkle. 1993. Influence of enrichment incubation time on the isolation of *Salmonella*. Avian Dis. **37**:884–887.
74. White, D.G., C.R. Hudson, J.J. Maurer, S. Ayers, S. Zhao, M.D. Lee, L.F. Bolton, T. Foley, and J. Sherwood. 2000. Characterization of chloramphenicol and florfenicol resistance in bovine pathogenic *Escherichia coli*. J. Clin. Microbiol. **38**:4593–4598.
75. Yaron, S. and K.R. Matthews. 2002. A reverse transcriptase-polymerase chain reaction assay for detection of viable *Escherichia coli* O157:H7: Investigation of specific target genes. J. Appl. Microbiol. **92**:633–640.

CHAPTER 3

Sample Preparation for PCR

Margie D. Lee[1,2*], DVM, Ph.D. and Amanda Fairchild[1], M.S.

Poultry Diagnostic & Research Center[1], College of Veterinary Medicine,
The University of Georgia, Athens, GA 30602
and
Center for Food Safety[2], College of Agriculture and Environmental Sciences,
The University of Georgia, Griffin, GA 30223

Introduction
How Do You Get Started?
What Conditions Affect the Success of the PCR?
What Are PCR Inhibitors?
Potential Solutions to the Challenges of Using PCR to Detect Pathogens in Foods
References

INTRODUCTION

Adding PCR-detection to a laboratory's repertoire of tools can improve sample turn-around time and accuracy. Yet, PCR is not a universal solution for pathogen detection problems. For pathogens that are rapidly growing and contaminate foods in high numbers, culture onto selective and differential media may actually be more rapid and cost-effective than PCR. However, PCR can greatly improve turn-around time in instances of slow-growing pathogens and can improve detection of pathogens present at low concentrations. Nevertheless, like any other protocol, correct preparation of the samples is key to PCR's success. Figure 3.1 shows a sample processing strategy for PCR. Specific steps in the processing strategy will vary depending on pathogen and foodstuff. Prior research may have shown whether pathogen amplification steps, such as enrichment culture, are needed prior to PCR, so check the published literature for relevant protocols.

HOW DO YOU GET STARTED?

Sample preparation serves several functions for PCR detection (14, 19). It initially decreases sample volume and concentrates the PCR template into a workable volume. The first challenge in choosing a good sample preparation protocol is to know whether the pathogen contaminates food at high levels or whether it

*Corresponding author. Phone: (706) 583-0797; FAX: (706) 542-5630; e-mail: leem@vet.uga.edu.

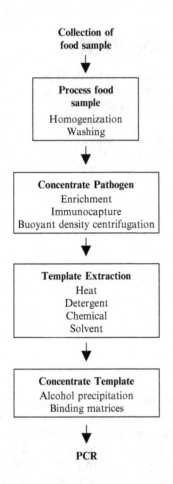

Figure 3.1. Sample preparation for PCR detection of foodborne pathogens.

will be necessary to amplify the bacteria with an enrichment culture. For example, very few *Listeria* cells may be present on a slice of deli meat but these few bacteria may be enough to cause serious illness in a pregnant woman. It may be impossible to directly collect a few bacterial cells and detect them using PCR. An enrichment culture can amplify the bacterial cells and the PCR can detect the bacteria in the enrichment broth more rapidly than they can be identified using standard bacteriological methods. In this instance PCR can aid in the rapid detection of *Listeria* and the sample preparation protocol will include performing the enrichment culture, collection of bacteria from the enrichment broth, extraction of DNA from the bacterial cells and then performing the PCR test.

Once the pathogen is collected, PCR template must be prepared from its DNA (or RNA). The first step in preparing template from a pathogen requires lysis (rupture) of the cells (or viruses) to release the nucleic acids (DNA and RNA). Specific organisms may require specific protocols for efficient template extraction. For example, there are a few basic approaches to extracting nucleic acids from bacteria but their effectiveness depends on several features of the bacterial cell wall. Gram-negative bacteria lack a thick cell wall, thus heat or detergent can lyse the cells. Many of the published protocols for *E. coli, Salmonella,* and *Campylobacter* use this approach for lysis of the cells (see Table 3.1 for applications). Gram-positive bacteria have a thick cell wall that must be removed or disrupted in order to lyse the cells. Lysozyme (plus lysostaphin for *Staphylococcus*) digestion is commonly used, prior to detergent treatment, for nucleic acid extraction from gram-positive and gram-negative bacteria. A third method of bacterial cell lysis involves a high salt/chemical lysis with guanidium salts. This method is most commonly used for gram-positive bacteria but will work for gram-negative cells as well. Solvent extraction of nucleic acids, with organic solvents such as ether, can be used for bacteria, viruses, and protozoa. Commercial PCR detection kits will incorporate one or some derivation of these methods, but the methods have to be optimized for the specific organism.

Table 3.1. Sample preparation of foods for PCR

Food Category (Challenges)	Sample	Method to Concentrate Pathogen	DNA (RNA) Extraction method	Pathogens (Reference)
Dairy PCR inhibitors (fat, protein, calcium, chelators), dead cells, low numbers of pathogen cells, other bacteria	Skim milk, pasteurized milk, dry milk, hard and soft cheese, reconstituted whey powder	Differential centrifugation or none	Solvent-based nucleic acid extraction or guanidinium isothiocyanate extraction	*E. coli O157* (16) *Listeria* (12) *Staphylococcus, Yersinia* (20) *Campylobacter* (24)
	Raw milk	Centrifugation	Boiled cells with Chelex-100 removal of inhibitors; *Tth* polymerase improved sensitivity	*Staphylococcus* (10)
	Raw milk	Enrichment and centrifugation	Commercial kits	*Salmonella* (5)
	Soft cheese	None	Detergent lysis with NaI extraction of DNA	*Listeria* (15)
Meat and poultry rinses PCR inhibitors (fat, protein, collagen, blood), small numbers of bacteria,	Chicken carcass rinses, red meat	Enrichment and centrifugation	Commercial kits	*Listeria Salmonella E. coli* (4, 5)
	Homogenates of chicken skin, whole chicken leg, chicken sausages, turkey leg meat, ground beef, mince meat, beef, pork	Buoyant density centrifugation	Guanidinium isothiocyanate and detergent extraction	*Campylobacter* (24)
	Raw whole chicken rinses	Buoyant density centrifugation and enrichment culture	Boiled cells	*Campylobacter* (27)
	Chicken and turkey muscle, skin, internal organs; raw carcasses	Enrichment	Multiple methods firmed including boiled cells, alkaline lysis, and commercial kits	*Salmonella* (6)

Continued

Table 3.1. Sample preparation of foods for PCR—*cont'd*

Food Category (Challenges)	Sample	Method to Concentrate Pathogen	DNA (RNA) Extraction method	Pathogens (Reference)
Meat and poultry rinses	Ground beef	Buoyant density centrifugation, immunoma-gnetic separation, enrichment	Boiled cells or Chelex-extraction	*E. coli* (25)
	Ham	Immuno-magnetic separation	Lysozyme and detergent extraction	*Listeria* (8)
	Minced Pork meat, raw whole pork leg	Enrichment and buoyant density centrifugation,	Commercial extraction buffer and heat	*Yersinia* (13)
	Ground pork	Enrichment	Chelex resin-based commercial kit	*Yersinia* (26)
	Sausage and meat rolls (Korean ethnic foods)	Homogeni-zation of food then filtration and centrifugation	Commercially available kits but increased Mg++ levels in samples	*Clostridium* (11)
	Deli meats: ham, turkey, roast beef	None	Commercial extraction solution	*Norwalk-like virus Hepatitis A virus* (22)
Seafood PCR inhibitors (phenolics, cresols, aldehydes, proteins, fats), low numbers of bacteria	Smoked salmon	Homogeni-zation of food	Detergent extraction and Tween 20 facilitator; PCR inhibitors removed by solvent extraction or column purification	*Listeria* (23)
	Fish cakes, fish pudding, peeled frozen shrimp, salted herring, marinated and sliced coalfish in oil	Enrichment	Detergent and boiling for extraction	*Listeria* (1)

	Shellfish: muscles and oysters	Homogenization of food then high-speed centrifugation	Guanidinium thiocyanate and silica purification	*Norwalk-like virus, Adenovirus Enterovirus, Hepatitis A virus* (7)
	Raw Oysters	Homogenization then buoyant density centrifugation	Commercial kit	*Norovirus* (17)
Produce PCR inhibitors (chelators), few bacteria	Whole raspberries	Column filtration and centrifugation	Commercial kit (FTA filter)	Protozoa (18)
	Lettuce	Homogenization, centrifugation, and precipitation with polyethylene glycol	Commercial kits	*Hepatitis A virus Norwalk virus* (21)

The next step in sample processing is to concentrate the template and reduce the concentration of PCR inhibitors. The specific approach will vary depending on whether DNA or RNA is desired as template and the chemical composition of the PCR inhibitors present in the sample. Phenol/chloroform extraction steps can reduce protein and lipid inhibitors. Other chemical inhibitors can be diluted by washing bound (silica beads or column matrices) or precipitated (ethanol or propanol) nucleic acids. An effective protocol for removing inhibitors must be developed for each specific food. Then the specific PCR can be performed for the pathogens of interest.

WHAT CONDITIONS AFFECT THE SUCCESS OF THE PCR?

The purpose of PCR is to detect the organism's specific nucleic acids in the sample so that time-consuming biochemical and immunological assays are not needed. PCR causes the synthesis of DNA using an enzymatic reaction that cycles over and over due to the temperature cycling of the thermal cycler. Enzymes, including the polymerases that are used in PCR, must have specific chemical conditions in order to do work effectively. One of the major concepts of PCR is that the polymerase can exponentially increase the amount of DNA in the sample because of the temperature cycling (9). However, if the conditions are not optimal, the polymerase may not be able to synthesize enough DNA for the reaction to be detected as positive; these are called "false-negative" reactions. There are many situations where the PCR reaction can be suboptimal and produce false negative results. Incorrect primers, buffer composition, cation (Mg++) concentration, nucleotide concentration (dNTPs), the wrong annealing temperature, extension cycles that are too brief, and incorrect template can cause the reaction to be falsely positive or negative. Always include two negative controls: a different organism's DNA and a control with no DNA template. These will help you determine the specificity of your PCR and whether you have sample contamination. In addition, always include a positive control with DNA template that you know will amplify in the PCR. These controls can help identify the problem when the PCR is inhibited. For example, different polymerases need different cations in the buffer in order to synthesize DNA. Taq polymerase uses magnesium (Mg++) therefore too little Mg++ in the master mix will result in a negative PCR reaction. The nucleotide concentration is important as well; they will chelate the Mg++ if you use too much of the dNTP mix. However, too much Mg++ will also result in a false negative reaction because there is a narrow window of effectiveness for the PCR to work. Every PCR reaction must be tested to determine the optimal concentration of Mg++ for the specific primers, buffer, and cycling temperatures. A similar situation also exists for the template concentration, too much or too little will result in a false negative reaction. You should optimize the PCR, by running different concentrations of cation and template to find the concentrations that will produce the amplicon of the correct size (or melting temperature if you are using real-time PCR). If you are setting up a new PCR that you found in published literature, do not just assume that the published conditions will work for you. If you get into the habit of optimizing your PCR reaction conditions, you will seldom have a problem with your routine PCRs that you cannot quickly solve.

WHAT ARE PCR INHIBITORS?

The PCR reaction can be inhibited when substances bind (chelate) or degrade a component in the reaction and prevent it from participating in the synthesis of DNA (9, 28). These substances are called "PCR inhibitors" and include chelators of cations and substances that bind or degrade the polymerase or the DNA template. When pathogens are grown to high levels in culture, PCR template can often be made directly by chemical or enzymatic lysis of the organism's cells. For example, DNA can easily be extracted from gram-negative bacteria by boiling the cells in water and using the boiled lysate as template. One important caveat, most enrichment broths and selective agars contain substances that inhibit PCR so it will be important to wash the cells collected from an enrichment or agar plate. You can do this with bacteria by pelleting the cells using centrifugation, removing the liquid and resuspending the cells in saline or water for the DNA extraction.

The real challenge is to isolate the pathogen and/or its DNA directly from a food matrix. The great variety of foodstuffs complicates any quest to produce a single sample preparation protocol that will work for every application. Unique PCR inhibitors are found in just about any food type including meat, milk, cheese, produce, and spices (28). Many of these have not been identified but some are known substances. For example, milk contains high levels of cations (Ca++), proteases, nucleases, fatty acids, and DNA. In addition, heme, bile salts, fatty acids, antibody, and collagen are PCR inhibitors that may be present in meat or liver. The inhibitors have variable effects on the PCR reaction but in general, they will make it more difficult to detect low numbers of bacterial cells or viruses. A good sample preparation protocol will focus on collecting the pathogen, removing the inhibitors present in the foodstuff (or culture medium) and concentrating the template for PCR. In addition, use of a polymerase that is less susceptible to the effects of inhibitory substances is a possible solution to some PCR problems. For example a number of the newer polymerases, such as *Tfl* and *rTth*, are more reliable than *Taq* polymerase when using PCR template made from meat or cheeses (2). Moreover, the activity of the polymerases, in the presence of inhibitors, can be improved with the use of some facilitators such as bovine serum albumin (BSA), dimethyl sulfoxide (DMSO), Tween 20 and betaine (2, 3, 19, 28). If you are trying to adapt a published PCR to a different food type, you may have to consider adding a facilitator or using a different polymerase to enhance sensitivity of the reaction.

POTENTIAL SOLUTIONS TO THE CHALLENGES OF USING PCR TO DETECT PATHOGENS IN FOODS

Foods differ greatly in their composition. The presence of fats, proteins, enzymes, chemical additives, fiber, and bacteria as well as ranges of pH influence your ability to isolate the organism, its nucleic acids, and amplify its nucleic acids using PCR. In addition, nonpathogenic organisms present in fermented foods, the contaminating soil and manure organisms present on produce, and fecal contamination of meat will produce competing DNA that may reduce the sensitivity of the PCR reaction. Unless you are experienced in developing PCR reactions, you may not want to solve all of these problems yourself. Use protocols developed and validated by reputable labs. Note specific steps in the protocol. Are enrichment steps such as immunomagnetic capture or enrichment culture needed to collect the organism from the sample? How are the organisms collected from the sample or the enrichment? How is the DNA extracted from the organism? What are the specific conditions of the PCR reaction? How will you detect the PCR amplicon? Do you have a positive control organism (or template) for the reaction? Once you have dissected out these important components from the publication or protocol, you can determine, which components can be modified for your specific needs.

Table 3.1 illustrates some of the challenges and solutions for PCR detection of pathogens that contaminate foods. If the challenges are acknowledged, then possible solutions become feasible. If the pathogen contaminates the food in low numbers, then the pathogen must be amplified in some way. The important thing to know is whether the PCR can detect very low numbers of bacteria. Theoretically PCR can detect 1 pathogen in the reaction. Yet realistically, because of PCR inhibitors and other factors, you will usually need a substantial amount of template, from a few hundred to thousands of pathogen cells or viruses, in the PCR reaction in order to reliably detect the presence of the pathogen. In addition, if you consider that some food samples may only contain a few hundred pathogen cells per gram of food, the need for an enrichment step becomes apparent. Figure 3.2 shows how common pathogen amplification methods work to concentrate the pathogen in a volume that can be used for the next steps in the sample preparation . Enrichment culture is commonly used to amplify the bacterial numbers although immunological capture can theoretically be used for bacteria, viruses, and protozoa. Nevertheless, immunocapture requires the availability of an antibody that is specific for the pathogen. An immunocapture system works by binding one end of the antibody to a handling apparatus (such as magnetic beads in immunomagnetic separation) then exposing the other end of the antibody to the contaminated food. If the pathogen is present in the food, many of the cells or virus particles will be bound to the antibody. This means that the immunocapture system can concentrate the pathogen onto the magnetic beads, which can then be used for enrichment or directly processed in a DNA extraction for PCR. Similarly, pathogens can be isolated from liquid samples by using centrifugation protocols that either float the bacteria or virus particles out of the sample (buoyant density centrifugation) or pellet the cells in the tube. Filtration of liquid samples may be effective for

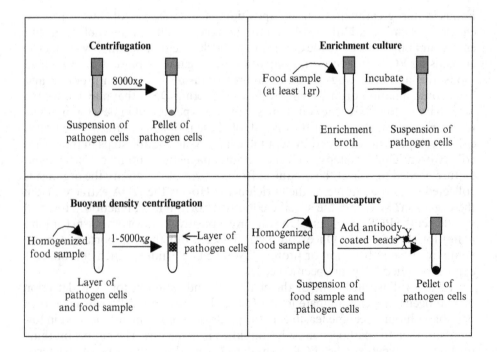

Figure 3.2. Approaches to concentrate or amplify pathogens.

concentrating pathogens that pass through a 0.45 micron filter (such as *Campylobacter* and virus particles). For some pathogens where the infectious dose is very low, more than one concentration step may be needed in order to amplify the pathogen to a detectable level. Many studies have addressed these detection issues in developing PCRs for specific applications (Table 3.1).

REFERENCES

1. Agersborg, A., R. Dahl, and I. Martinez. 1997. Sample preparation and DNA extraction procedures for polymerase chain reaction identification of *Listeria monocytogenes* in seafoods. Int. J. Food Microbiol. **35**:275–280.
2. Al-Soud, W.A. and P. Radstrom. 1998. Capacity of nine thermostable DNA polymerases to mediate DNA amplification in the presence of PCR-inhibiting samples. Appl. Environ. Microbiol. **64**:3748–3753.
3. Al-Soud, W.A. and P. Radstrom. 2000. Effects of amplification facilitators on diagnostic PCR in the presence of blood, feces, and meat. J. Clin. Microbiol. **38**:4463–4470.
4. Bhaduri, S. and B. Cottrell. 2001. Sample preparation methods for PCR detection of *Escherichia coli* O157:H7, *Salmonella typhimurium*, and *Listeria monocytogenes* on beef chuck shoulder using a single enrichment medium. Mol. Cell. Probes **15**:267–274
5. Chen, S., A. Yee, M. Griffiths, C. Larkin, C.T. Yamashiro, R. Behari, C. Paszko-Kolva, K. Rahn, and S.A. De Grandis. 1997. The evaluation of a fluorogenic polymerase chain reaction assay for the detection of *Salmonella* species in food commodities. Int. J. Food Microbiol. **35**:239–250.
6. De Medici, D., L. Croci, E. Delibato, S. Di Pasquale, E. Filetici, and L. Toti. 2003. Evaluation of DNA extraction methods for use in combination with SYBR Green I Real-Time PCR to detect *Salmonella enterica* Serotype Enteritidis in poultry. Appl. Environ. Microbiol. **69**:3456–3461.
7. Formiga-Cruz, M., G. Tofiño-Quesada, S. Bofill-Mas, D.N. Lees, K. Henshilwood, A.K. Allard, A.-C. Conden-Hansson, B.E. Hernroth, A. Vantarakis, A. Tsibouxi, M. Papapetropoulou, M.D. Furones, and R. Girones. 2002. Distribution of human virus contamination in shellfish from different growing areas in Greece, Spain, Sweden, and the United Kingdom Appl. Environ. Microbiol. **68**:5990–5998.
8. Hudson, J.A., R.J. Lake, M.G. Savill, P. Scholes and R.E. McCormick. 2001. Rapid detection of *Listeria monocytogenes* in ham samples using immunomagnetic separation followed by polymerase chain reaction. J. Appl. Microbiol. **90**:614–621.
9. Kainz, P. 2000. The PCR plateau phase—Towards an understanding of its limitations. Biochimica et Biophysica Acta **1494**:23–27.
10. Kim, C.H., M. Khan, D.E. Morin, W.L. Hurley, D.N. Tripathy, M. Kehrli Jr., A.O. Oluoch, and I. Kakoma. 2001. Optimization of the PCR for detection of *Staphylococcus aureus nuc* gene in bovine milk. J. Dairy Sci. **84**:74–83.
11. Kim, S., R.G. Labbe, and S. Ryu. 2000. Inhibitory effects of collagen on the PCR for detection of *Clostridium perfringens*. Appl. Environ. Microbiol. **66**:1213–1215.
12. Koo, K. and L.A. Jaykus. 2003. Detection of *Listeria monocytogenes* from a model food by fluorescence resonance energy transfer-based PCR with an asymmetric fluorogenic probe set. Appl. Environ. Microbiol. **69**:1082–1088.
13. Lantz, P.G., Y. Knutsson, Y. Blixt, W.A. Al-Soud, E. Borch, and P. Radstrom. 1998. Detection of pathogenic *Yersinia enterocolitica* in enrichment media and pork by a

multiplex PCR: A study of sample preparation and PCR-inhibitory components. Int. J. Food Microbiol. **45**:93–105.

14. Lantz, P.G., W.A. Al-Soud, Ri. Knutsson, B. Hahn-Hagerdal, and P. Radstrom. 2000. Biotechnical use of polymerase chain reaction for microbiological analysis of biological samples. Biotechnol. Ann. Rev. **5**:130.

15. Makino, S.I., Y. Okada, and T. Maruyama. 1995. A new method for direct detection of *Listeria monocytogenes* from foods by PCR. Appl. Environ. Microbiol. **61**:3745–3747.

16. McKillip, J.L., L.A. Jaykus, and M.A. Drake. 2000. A comparison of methods for the detection of *Escherichia coli* O157:H7 from artificially contaminated dairy products using PCR. J. Appl. Microbiol. **89**:49–55

17. Nishida,T., H. Kimura, M. Saitoh, M. Shinohara, M. Kato, S. Fukuda, T. Munemura, T. Mikami, A. Kawamoto, M. Akiyama, Y. Kato, K. Nishi, K. Kozawa, and O. Nishio. 2003. Detection, quantitation, and phylogenetic analysis of Noroviruses in Japanese oysters. Appl. Environ. Microbiol. **69**:5782–5786.

18. Orlandi, P.A., and K.A. Lample. 2000. Extraction-free, filter-based template preparation for rapid and sensitive PCR detection of pathogenic parasitic protozoa. J. Clin. Microbiol. **38**:2271–2277.

19. Radstrom, P., R. Knutsson, P. Wolffs, M. Lovenklev, and C. Lofstrom. 2004. Pre-PCR processing strategies to generate PCR-compatible samples. Mol. Biotechnol. **26**:133–146.

20. Ramesh, A., B.P. Padmapriya, A. Chandrashekar, and M.C. Varadaraj. 2002. Application of a convenient DNA extraction method and multiplex PCR for the direct detection of *Staphylococcus aureus* and *Yersinia enterocolitica* in milk samples Mol. Cell. Probes **16**:307–314.

21. Sair, A.I., D.H. D'Souza, C.L. Moe, and L.A. Jaykus. 2002. Improved detection of human enteric viruses in foods by RT-PCR. J. Vir. Meth. **100**:57–69.

22. Schwab, K.J., F.H. Neill, R.L. Fankhauser, N.A. Daniels, S.S. Monroe, D.A. Bergmire-Sweat, M.K. Estes, and R.L. Atmari. 2000. Development of methods to detect "Norwalk-Like Viruses" (NLVs) and Hepatitis A Virus in delicatessen foods: Application to a foodborne NLV outbreak. Appl. Environ. Microbiol. **66**:213–218.

23. Simon, K.C., D.I. Gray, and N. Coo. 1996. DNA extraction and PCR Methods for the detection of *Listeria monocytogenes* in cold-smoked salmon. Appl. Environ. Microbiol. **62**:822–824.

24. Uyttendaele, M., J. Debevere, and R. Lindqvist. 1999. Evaluation of buoyant density centrifugation as a sample preparation method for NASBA-ELGA detection of *Campylobacter jejuni* in foods. Food Microbiol. **16**:575–582.

25. Uyttendaele, M., S. van Boxstael, and J. Debevere. 1999. PCR assay for detection of the *E. coli* O157:H7 *eae*-gene and effect of the sample preparation method on PCR detection of heat-killed *E. coli* O157:H7 in ground beef. Int. J. Food Microbiol. **52**:85–95.

26. Vishnubhatla, A., D.Y.C. Fung, R.D. Oberst, M.P. Hays, T.G. Nagaraja, and S.J.A. Flood. 2000. Rapid 5′ nuclease (TaqMan) assay for detection of virulent strains of *Yersinia enterocolitica*. Appl. Environ. Microbiol. **66**:4131–4135.

27. Wang, H., J.M. Farber, N. Malik, and G. Sanders. 1999. Improved PCR detection of *Campylobacter jejuni* from chicken rinses by a simple sample preparation procedure Int. J. Food Microbiol. **52**:39–45.

28. Wilson, I.G. 1997. Minireview: Inhibition and facilitation of nucleic acid amplification. Appl. Environ. Microbiol. **63**:3741–3751.

CHAPTER 4

Making PCR a Normal Routine of the Food Microbiology Lab

Susan Sanchez, Ph.D.

Athens Diagnostic Laboratory, and the Department of Infectious Diseases, College of Veterinary Medicine, The University of Georgia, Athens, GA 30602

Introduction
Setting up Your Laboratory for PCR
  Physical Set-up of PCR
  Personnel
  Real-Time vs. Standard Format PCR
  Equipment
  Reagents and Disposables
  Quality Control and Quality Assurance
  Where to Locate Vendors
Noncommercial Tests for Foodborne Pathogens
  Validation
  Standardization
Available Commercial PCR Tests for Foodborne Pathogens
  Real Time PCR
    BAX Dupont-Qualicon
    Food-proof Roche
    IQ-Check BioRad
  PCR-ELISA
    Probelia BioRad
    AnDiaTec Salmonella sp. PCR-ELISA
References

INTRODUCTION

In food microbiology, polymerase chain reaction (PCR) should not be considered a substitute for conventional microbiology techniques. The rationale of employing PCR technology, as well as any other molecular diagnostic technique, should be founded on the following key consideration: (1) simplicity, (2) throughput, (3) cost, (4) speed, and (5) appropriateness (37). Conventional bench microbiology is often considered to be less technically demanding than polymerase chain reaction, however, in reality, the techniques of PCR are easier to master and usually requires less time to achieve competence than conventional microbiology. Experience in our laboratory shows that personnel of varied technical and educational backgrounds, and absolutely no training in microbiology, can master polymerase chain reaction in no more than a week. By contrast, competence in conventional microbiology techniques and their interpretation requires significant training and experience. Polymerase chain reaction can be mastered in less than a week with very little or no previous training in microbiology, whereas conventional microbiological techniques will require several weeks of training and a background in microbiology. A technique, as simple as PCR, can be applied in more laboratories and it is more amenable to field applications, which could allow for data collection right at the food-processing plant. The food microbiologists' constant aim is to make our food supply safer; this is only achieved with large-scale screening of foodstuffs. Polymerase chain reaction is a test process that allows for high throughput and is amenable to automation. Over the past few years, molecular reagents have become more affordable, and easier to obtain, with longer shelve life (37). Naturally, the cost and accessibility varies greatly with location. Except for primers and probes, most PCR reagents can be used and shared among different PCR tests, as well as for other molecular-based techniques used in food microbiology (e.g., strain typing) (37). The decision to use PCR needs to be made by the laboratory based on clients' requirements for quick turnaround, reliability, and confidence in the laboratory for correctly reporting results. Real-time PCR allows not only the detection of suspect pathogen in food but its enumeration as well (36, 37, 38). These molecular tools are essential to a contemporary food laboratory. Although the cost to equip a laboratory is high, PCR complements and enhances the traditional microbiological methods; by increasing speed, sensitivity, and, specificity for detecting pathogens in foods. Polymerase chain reaction can be performed rapidly in the field, and limits the number of cultures and isolations to the few samples identified as positive by PCR (28). This reduces labor, conserves resources, and holds down costs.

SETTING UP YOUR LABORATORY FOR PCR

The PCR is a very sensitive, exceptionally powerful, and relatively simple method that can unlock the door to many a genetic mystery after a few rounds of cycling temperatures. However, PCR can return false positives or negatives if care is not taken in the proper setup, standardization, and implementation of quality controls. Careful planning needs to be given to standardizing a routine protocol(s) for processing and testing of samples by PCR. This includes: the physical setup of the PCR laboratory; quality control; storage and selection of reagents; deciding which PCR detection platform to use; and finally, a critical component—well-trained personnel. With good laboratory practices, the diagnostic laboratory avoids the possible pitfalls with PCR that lead to erroneous reporting of laboratory results.

Physical Setup of PCR. Avoiding sources of PCR contamination is paramount when dedicating laboratory space to PCR. Ideally there should be 4 to 5 distinctly separate rooms or areas in the laboratory for: (1) sample preparation, which includes sample enrichment, clean reagent preparation, and DNA extraction; (2) PCR setup; (3) thermocycler; and (4) detection of PCR products or amplicons by agarose gel electrophoresis, or enzyme-linked immunosorbent assay (ELISA). Realistically, having this much room may not be possible in existing food microbiology laboratory, although it should be kept in mind if future renovations are planned, or a new building is being designed. In most instances, this physical separation can be accomplished by the development of one-way traffic flow (Fig. 1) within a laboratory, assigning particular areas to each step in the process. This will hopefully prevent cross-contamination from sample to sample and PCR to PCR. The assignment of areas and/or rooms needs to be intuitive and easy to follow so that it does not influence the speed and normal work flow of the laboratory. Ideally, personnel will start the day in the clean area and move to those areas where there is higher risk of aerosols that could be transported to clean areas and become sources of PCR contamination. In those laboratories that process a very large number of samples, dedicated personnel to each area is a good idea with rotation of these people to a different area every week to avoid boredom and maintain full competence in all aspects of the PCR protocol. Laboratory coats or scrubs should be worn always, but these garments need to remain in each area except when cleaned. DNA present on a bench top can easily contaminate a sleeve. If a technologist later uses the same laboratory coat, then the jacket could serve as the source of PCR carryover contamination when the individual sets up the next PCR reaction. Gloves should be used at all times and should never be worn in the different PCR areas. *The sample preparation area is one of the most important areas in the laboratory and it should be divided in three individual areas for: (1) food sample preparation, (2) clean reagent preparation for DNA extraction, where no sample or template should ever be present; and (3) processing and extraction of DNA from foods or enrichments.* This final step should be conducted in a biological safety cabinet to avoid sample cross-contamination. The technician can inadvertently contaminate sample(s), from which PCR template is prepared by

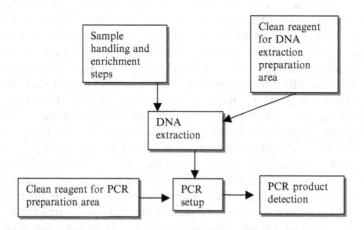

Figure 4.1. Chart delineating the workflow in the PCR laboratory. Some of the PCR work areas need to be physically separated in order to minimize the potential for PCR carryover contamination, or sample cross-contamination. DNA template or PCR product should never enter the clean reagent preparation area.

touching a small droplet from an enrichment broth, positive for the microbe in question, and subsequently can transfer this to another tube containing a different sample, by touching the second sample with contaminated gloved hand. This cross-contamination is insidious as it will not be detected by our PCR controls and, even if contamination is suspected, retesting of the contaminated sample template will only yield the same spurious false-positive results. The sample will have to be reextracted, adding time and cost to the process. *The PCR can be exquisitively sensitive, even detecting a few cells that may cross-contaminate a "negative" sample.* To avoid this scenario, care should be taken in cleaning the outside of the sample tubes with a disinfectant to kill bacteria and DNAse, or 10% bleach solution to destroy any contaminating DNA. It is also recommended that the cabinet and the pipettes are wiped with bleach solution to eliminate sample cross-contamination, and subsequently wiped clean with ethanol or water to avoid corrosion by the bleach, before the next DNA extractions (13). Contamination by aerosols can also be minimized if supernatants are aspirated and not decanted. When large volumes of sample need to be handled, disposable individually wrapped sterile pipettes should be used. DNA extraction from already enriched samples is one of the PCR steps that is amiable to automation, which reduces the risk for intersample cross-contamination and sample exchange (30, 49). *The PCR setup* and *thermoycler area* can be physically separated but they can also be placed together as long as there is a biosafety cabinet in this area, or a PCR box for the PCR setup. "PCR clean hoods" are relatively inexpensive, sit on a table or bench and they are easy to clean. Their relatively small size allows for their placement in most laboratories. PCR clean hoods also have germicidal ultraviolet (UV) lights, which aid in maintaining sample purity by denaturing DNA contaminants (13). The use of UV light to decontaminate the PCR setup area, where reagents are openly handled can minimize the risk of carryover contamination (40).

The *detection area* itself, where tubes or capillaries are opened and loaded into agaraose gels or ELISA microplates, can serve as a source of PCR contamination. PCR can amplify DNA from very small amounts of template to large quantities of amplicon, which once aerosolized, aspirated, or spilled, can easily contaminate surfaces. The glass capillaries used with the hot-air thermocyclers can break within their carrier during transport to and from thermocycler and detection area, or required glasscutter to free it from its block. Breakage can result in contamination of this block with amplicon. Therefore, pipettes as well as racks and carriers, used to transport samples from PCR setup to thermocycler to detection area and back, can provide an opportunity for contamination of the next scheduled PCR run. This issue is largely avoided with thin-walled microcentrifuge tubes and other specialized microtubes for real-time PCR. These tubes are not likely to break and contaminate the holders, the only piece of equipment that will go back in the PCR setup area for future use. Barrier tips can prevent inadvertent suction of fluids into the barrel of pippettors. Also a separate set of pipettors also eliminates pipettors as vehicles for introducing PCR carryover contamination during setup. Real-time PCR avoids carryover contamination due to the "real-time" detection of the amplicons as they are produced, and thus eliminates the necessity from ever having to open the tube once the PCR reaction has been set up to run.

With each additional PCR run, the risk of false-positives increases for that test. Therefore, minimizing the risk using the best work flow and work practices possible is paramount. The size and the number of working areas that can be used simultaneously at one given time will have to be anticipated based on the number of samples that the laboratory plans to process daily, and how many personnel will be working in each area at the same time.

Personnel. The PCR is simpler than traditional microbiology, not only regarding the procedure itself, but also because mastering this new technology takes less time than required to learn conventional microbiology. It can be mastered in less than a week with very little or no previous training in microbiology, whereas conventional techniques will require several weeks of training and a background in microbiology. Competent conventional microbiologists can be easily trained to perform PCR and will quickly discover the advantages of this methodology. It can provide rapid same or next day results and, as an initial screen, it can streamline culture and isolations to the few presumptively positive samples (Fig. 2). If we only check PCR-positive samples, we decrease our workload. For the laboratory, this reduces labor, resources, and finally cost. Labor costs rise exponentially as the number of microorganisms to be ruled out increases when comparing PCR to bacterial culture methods (47). Multiplex PCR and microarrays discussed in Chapter 1, reduces these pathogen screens to a single test (12, 26, 41, 46). Unfortunately, several of the commercial tests discussed later in this chapter are detection tests for single foodborne pathogen, and single multipathogen-detection systems are in their infancy. While detection and final confirmation of foodborne pathogen may take up to 6 days with standard microbiology protocols, PCR can provide clientele with preliminary results while the microbiology lab continues to work up the submission. Laboratory supervisory personnel need to be knowledgeable in interpreting PCR results, recognizing and correcting problems as they develop, and in explaining the significance of PCR results to the clientele. Chapter 2 discusses interpretation of PCR results. Personnel with enough molecular training needs monitor result output and keep track of the number of positive samples, to request the retesting of positive samples suspected of being cross-contaminated. The personnel should also keep abreast of the literature regarding PCR to discover improvements and problems with current tests and identify new primers, PCR assays, or methodology available for foodborne pathogens confronting the food industry.

REAL-TIME VS. STANDARD FORMAT PCR

The PCR amplicons are detected either in "real time" as they are synthesized, or following PCR run on an agarose gel, or in an ELISA microplate. The suitability of either real-time or standard format PCR will differ according to the laboratory's needs, resources, and expertise.

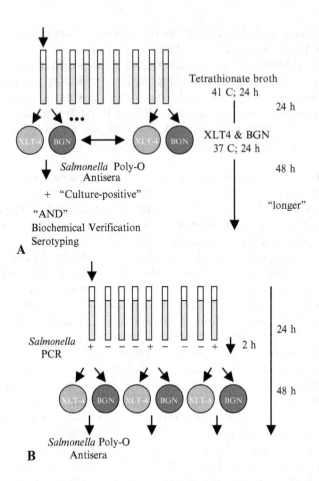

Figure 4.2. PCR as tool to identify samples or enrichments, which need to be processed further. (A) Steps and time required to identify sample(s) contaminated with PCR. (B) Incorporation of PCR into screen of samples, or culture enrichments to identify samples requiring additional culture/platings to identify the foodborne pathogen, in this example *Salmonella enterica*.

Amplicon detection can be done without agarose gels. In PCR-ELISA, the amplicon hybridizes with a specific, internal oligo probe tagged with biotin, which is subsequently bound and captured by streptavidin-coated ELISA microplate. As the amplicon was labeled with digoxygenin-tagged nucleotides during PCR, the bound amplicon-oligoprobe complex is detected colorometrically with antidigoxygenin antibody conjugated either to the enzymes alkaline phosphatase or horseradish peroxidases. Color change is subsequently recorded using an ELISA plate reader. With real-time PCR, a fluorimeter is built into the thermocycler to monitor fluorescence as the machine cycles through its denaturing, annealing, and extension step. As the double stranded amplicon is produced, a fluorescent dye, called SYBR Green, or specific probes labeled with a fluorescent dye bind to the PCR product and the detector measures the resulting fluorescence. The advantageous of either PCR-ELISA or real-time PCR over agarose gels in the detection of amplicons is their amendability to 96-well-microplate formats and automation (31, 43, 46). However, the need for agarose gels for detecting amplicons is not eliminated with either PCR detection format, as it still has a use in validation; the occasional trouble shooting, as problems arise (see Chapter 2); and implementation of real-time PCR or PCR-ELISA into the laboratory's routine.

Equipment. When talking about the laboratory adopting PCR, the first and in most cases the only piece of equipment that comes to mind to most readers is the thermocycler. Although, by far it is the most expensive individual piece of equipment, we should not lose sight of other expenses accrued in setting up the laboratory to perform PCR. Several sets of pipettes are needed for DNA extraction and PCR setup, a minimum of four sets. If the laboratory does not own enough biosafety cabinets, PCR clean hoods can be used. Electrophoresis equipment is required for amplicon detection in gels, the microwave oven needed for melting agarose, and enough $-20°C$ freezer space to keep all the necessary PCR reagents separated from the sample template to be tested. For PCR-ELISA, it will be necessary to also possess an ELISA reader, although this is an essential component already in place for any laboratory that does serology (46). Finally, the laboratory will need to record PCR results. For standard PCR format, where agarose gels are used to detect amplicons, the laboratory will need UV transilluminator and a camera to capture the gel image.

There are a number of factors that will influence which type of thermocycler is purchased. The overall cost of the machine itself might be a major deciding factor in a laboratory's ability to take advantage of PCR technology at all. A simple, hot-air, or conventional heating block thermocycler cost a few thousand dollars. However, the price can vary depending on user's needs and requirements regarding sample throughput (48 vs. 96), sample format (tube vs. 96-well microplate), and PCR reaction volume. The laboratory also needs to consider the cost of the PCR assay with regards to the individual reagents, disposables and ancillary components (e.g., wax beads or mineral oil) in selecting a thermocycler. For example, thermocyclers with heated lids allow for smaller PCR reaction volumes and eliminates the need for mineral oil in the PCR reaction. The speed of a 30-cycler program, 10 vs. 90 min, may be another factor in the purchase of a thermocycler. The laboratory may want to consider the benefits of a gradient thermocycler, which allows users to perform, simultaneously, several PCR tests that require different PCR cycle parameters. These gradient thermocyclers also have the advantage of identifying optimal-annealing temperature for a single or multiple PCR primer sets in a single run, due to the machine's ability to assign separate programs to each well. The price for real-time PCR technology jumps to between $30,000 and $140,000, depending on the components of the unit such as 96-well format and automation module. In selecting a thermocycler, the laboratory also needs to determine if the PCR will be an in-house validated assay or a commercial assay, as the latter will require specific PCR equipment. The following questions need to be addressed before purchasing a PCR thermocycler: Will there be many samples submitted on a regular basis or only periodically that require this type of technology? Will the samples only be a part of the usual submitted workload? Will the samples require detection of only one gene of interest, or are there various genes to identify, which would require different programs for the thermocycler? The PCR format chosen by the laboratory is a big deciding factor to determine the needs for equipment as we have seen before, because initial cost of equipment and applications of the equipment vary.

All instruments, from pipettes to freezers, must be calibrated and certified according to the respective institutional standard operating procedures, which in most cases are dictated by the institutions accreditation agency. For most institutions it is at least once a year. Most large equipment manufacturers offer yearly contracts for calibration of their equipment.

Reagents and Disposables. Reagents for the PCR consist of those used for the reaction that takes place in the thermocycler, and those used for the detection of amplicons. Purchasing of reagents should be done from a reputable company and molecular grade should be requested for any reagent that is going to be used for PCR. Primers and probes as well as *Taq* DNA polymerase can be contaminated with extraneous DNA. Top quality components should only be used and when handling these reagents, gloves should be worn as to prevent the introduction of metal ions, nucleases, or other contaminants to the PCR reaction (48). Water is an important component. Always use it as aliquoted, single use volumes that have been filtered (0.22 μm), and autoclaved. If money is not a problem, then ideally DNA-free and DNAse-free water should be purchased for PCR. We must not forget that the PCR reaction requires a DNA template, therefore; the required reagents for extraction or cell lysis need to be taken into account. Consumption and cost of reagents is related to the machine and the DNA extraction method that is used and the number of samples to be analyzed. Depending on the machine, volumes for each reaction can vary from 10 to 100 μl. The average volume used in PCR reactions is 25–50 μl. The smaller the PCR reaction volume, the cheaper the cost per PCR test is. Costs for *Taq* DNA polymerase, dNTPs, and other components of the PCR can turn PCR into an expensive venture for a laboratory very quickly. Lastly, one must consider the cost of labor for PCR setup as well as for sample preparation. It has been reported that the overall costs of the reagents and materials involved in identifying specific bacterial agents by PCR were 2 to 5 times higher than the costs involved with bacterial culture identification (17). As molecular reagents are continuously getting better in quality and longer in shelf life making them overall less expensive, this may not hold true. A good example of this is the reduction in cost of oligonucleotide synthesis.

In cost analysis, one must consider whether certain alternative traditional microbiological methods are feasible or practical with the laboratory's current resources and manpower. For example, *Salmonella* serotyping requires the laboratory to keep an extensive battery of antisera to identify the thousands of different *Salmonella* serovars. Considering the limited shelf-life of the necessary serotyping reagents, the laboratory's sample volume, and time it takes to identify the *Salmonella* serovar, a PCR-based approach might be a more time- and cost-effective approach (19, 21). Likewise, several of foodborne pathogens described in Chapters 6 and 7 are either recalcitrant to current culture, or isolation methods and PCR is more cost effective compared to the alternative isolation procedures. Where PCR trumps culture-based detection methods is that it can provide rapid preliminary results for making important decisions (15, 39, 41). For the laboratory, the decision as to which samples to work up further, and for the clientele, which lots to hold and which ones to ship.

The thermocycler and its throughput capacity will dictate the type of container used to hold PCR reaction: glass capillary tubes vs. thin-walled plastic; PCR microfuge tubes vs. 96-well microplate. Pipette tips must have filter barriers to avoid contamination of the inside of the pipette.

All newly prepared and purchased PCR reagents require quality control before use. This is necessary for any food microbiology laboratory to become successfully proficient at PCR.

Quality Control and Quality Assurance. Food microbiology laboratories are used to working under standard operating procedures (SOPs), as their laboratories are accredited to perform food microbiology. Each SOP not only contains the information on precisely how to conduct the procedure, but will also have information regarding when and how much quality control (QC) needs to be performed. A QC program usually requires all products and reagents (from DNA extraction to PCR) of each lot received to be tested to make sure they meet the same standard as previous lots so they can be utilized with confidence in the procedures. A good practice is to aliquot all reagents in single-use format after QC testing. Sometimes the testing of several aliquots is a good idea, especially if the laboratory does large volume of PCR routinely. Preparation of an aliquoted ready-to-use PCR master mix, that has been QC tested, is a good practice because this minimizes technician error resulting from miscalculation, or from forgetting to add a key component in the master mix. Commercially available kits should come with all reagents previously tested and this information if not in the packet insert should be available upon request. SOPs must include information regarding the required positive and negative amplification controls, as these will determine the stringency and accuracy of our PCR tests. The basic PCR controls are: DNA extraction controls, sample purposely spiked with the organism of interest, and another spiked with a different, unrelated organism; and PCR controls, pure, known amount of DNA from the organism of interest, and negative, no DNA, control. A QC for PCR can be made even more robust if an internal amplification control is included to check for the presence of PCR inhibitors in our samples (22, 23, 34). For more information about PCR inhibitors (see Chapters 3 and 4). Good SOPs and a good QC program will help minimize mistakes due to bad reagents and human error. Every laboratory should run a quality control program that is applicable and relevant for this methodology, and which is capable of detecting deficiencies at any level. This process can be made simpler most times by consulting with other labs performing similar PCR assays and purchasing reagents from suggested reputable vendors, although this will not eliminate the need for QC testing. SOPs are good training guides for new staff. Good record keeping is essential to any laboratory, where a bound laboratory notebook is kept with detailed and dated descriptions of protocols, reagents (including lot numbers, purchase, and expiration dates), controls, and results (32). Furthermore, a good quality assurance (QA) program at the institution ensures that there is compliance with SOPs for the PCR assays and consistency is achieved. Additional measures to consider are addressed in Chapter 2, and for a more thorough review of PCR laboratory setup, see *Methods in Molecular Medicine, Vol. 16: Clinical Applications of PCR* (29).

Where to Locate Vendors. For setting up your laboratory to do diagnostic PCR, you will need equipment and reagents to routinely perform PCR. Reagent-wise, you will need the enzyme, buffers, nucleotides, and barrier tips for dispensing reagents into thin-walled PCR tubes that contain the PCR reaction. In addition to this, you will need a source for custom synthesis of your oligonucleotides and probes needed for PCR. You will need to purchase agaraose gel electrophoresis and photo documentation equipment (film or digital-based), for analysis and documentation of conventional PCR results, as well as source for agarose, electrophoresis buffers, loading dye, and molecular weight (MW) standards. We have listed in Table 1, several sources for reagents and equipment listed above. *This list of companies is not an endorsement of the companies or their products;* rather, the table provides the reader an idea of what will be needed to implement PCR into food microbiology laboratory.

NONCOMMERCIAL TESTS FOR FOODBORNE PATHOGENS

Your laboratory after much consultation has decided that you need PCR as an additional method to evaluate your food samples. Now you need to make a decision which PCR format you want, provided you have a good physical infrastructure that will allow you to have the required physically separated areas in your laboratory, and you also have capable and trained personnel and a good QC program. Your laboratory has also decided that the use of a commercial test is not applicable, but instead the laboratory is going to use currently published primers, or even design for a much better PCR primer set. Published data by one laboratory can sometimes be difficult to reproduce due to the nature of the reagents, the variation in equipment, and the personnel training. Validation based on consensus criteria, detection limit, diagnostic accuracy (the degree of correspondence between the response obtained by the PCR method and the response obtained by the reference method on identical culture samples [AC = (PA + NA)/total number of samples; where PA = positive agreement; NA = negative agreement], diagnostic sensitivity, diagnostic specificity, and robustness, is a must for a successful microbiology laboratory (24, 25, 33). Your

Table 4.1. Molecular biology vendors of PCR

Vendor	Web Address	Product(s)
BIO-RAD	www.bio-rad.com	Thermocyclers, Electrophoresis apparatus, buffers, and MW standards, Photo documentation system
Dupont Qualicon	www.qualicon.com	PCR diagnostic tests (e.g., Bax *Salmonella*)
EPICENTRE	www.epicentre.com	PCR reagents, DNA cloning reagents
Fisher Scientific Co.	www.fishersci.com	PCR tubes, barrier tips, etc. Electrophoresis apparatus, Photo documentation, Agarose, Electro-phoresis buffer, Micropipettors, −20°C Freezer
Idaho Technology Inc.	www.idahotech.com	Thermocyclers, design and synthesis of primers and probes
Invitrogen	www.invitrogen.com	PCR reagents, PCR cloning vectors
MO BIO Laboratories Inc.	www.mobio.com	Nucleic acid extraction kits
Molecular Probes, Inc.	www.probes.com	Fluorescent dyes
Promega	www.promega.com	PCR reagents, PCR cloning vectors, MW standards, gel loading dye
Roche Applied Science	www.roche-applied-science.com	PCR reagents, Thermocycler
SeqWright	www.seqwright.com	DNA sequencing
Sigma-Genosys	www.sigma-genosys.com	Custom oligonucleotide synthesis
USA/Scientific Inc.	www.usascientific.com	PCR clean hood

laboratory will need to implement and then validate the PCR tests, bench-marking performance by demonstrating that the new method can generate results that are equal, or better, than those obtained by the current gold standard for detection. For a commercial PCR test that has already been validated, implementation is the only step required by your laboratory (34). One last point, when deciding which PCR format your laboratory is going to select, keep in mind the compatibility of the PCR tests chosen with the laboratory's current instrumentation and training. Molecular diagnostic tests that require new equipment, more laboratory space, and more training may not be the best choice. Therefore, a concerted effort at the initial planning stages should be made to foresee future demands.

Validation. There is no perfect PCR test and interlaboratory variation in performance of a PCR does occur. However, before diagnostic labs accept a PCR, there is a requirement for multilaboratory confirmation of the tests, specificity, sensitivity, and reproducibility. (See Chapter 2, discussion of validation.) Spiked samples as well regular samples should be used in the validation process as to mimic as much as possible everyday samples and situations. The extent your lab plays in the validation process of a PCR will be directly related to the target organism, food matrix, and previous work describing validation of PCR in peer-reviewed publications.

Currently, there is not a single harmonized validation protocol available. In 1999, the European Union (EU), through an initiative titled "the FOOD-PCR Project" (http://www.PCR.dk), set out to validate and standardize PCR for the detection of pathogenic bacteria in food using nonproprietary primers. The intended outcomes of this project were production of guidelines and kits for proficiency testing of different brands and types of thermocyclers, method for DNA extraction and purification, production of reference DNA material, and an online database containing validated PCR protocols. These protocols and results are available from their website. A further attempt has been made by the EU through the ISO/TC34 committee in collaboration with CEN/TC275, through the proposal EN ISO/FIDS 16140 "Microbiology of food and animal feeding stuffs—Part 42: Protocol for the validation of alternative methods" (1). Validation through this method seems more suited for commercially developed tests as the process is costly for nonproprietary, "home brew" PCR (34). Several commercial diagnostic tests have been validated in a very extensive manner and have been accredited by organizations on standards both at the international and national level (14, 20, 44). Current commercially available tests for the detection of foodborne pathogens will be reviewed later in this chapter.

Standardization. Standardization of PCR tests and extraction protocols at the national and international level will allow for accurate interlaboratory comparison. This can be achieved either with commercial tests, or with validated published primers. Standardization allows for fast implementation of tests, warranted accuracy, and detection limits, as well as known strength to allow for some variation in the tests procedure without giving misleading results. Standardization will also guarantee continued research and improvement of the PCR assay and protocols. Thus, the PCR test will fulfill its promise of being simple, high-yield, fast, appropriate, and even cheaper than the traditional culture (24, 34).

AVAILABLE COMMERCIAL PCR TESTS FOR FOODBORNE PATHOGENS

There is a wealth of ready-to-use PCR-based tests for the most common foodborne pathogens: *Salmonella* spp., *Listeria* sp. and *E. coli* O157:H7. The availability of commercial PCR tests for other foodborne pathogens is sparse and laboratory "home-brewed" PCR tests may be a better, if not the only, option. Chapters 5–7 describe several published PCR tests for bacterial, viral, and protozoal-foodborne pathogens. Several real-time and standard format PCR are commercially available. These two formats need to be studied carefully by each individual laboratory introducing PCR into their routine, chiefly to determine whether the equipment cost in comparison with potentially improved diagnostic ability is a worthwhile endeavor to pursue. One or more national and international agencies have validated commercial PCR tests described in this section (14, 20, 44). Their sensitivities and specificities are well known (14, 18, 20, 35, 41, 44, 45), and information is readily available form the products websites (www.andiatec.com; www.bio-rad.com; www.qualicon.com/bax.html; and www.roche-applied-science.com). Once a decision has been made on a PCR-detection format, then the choice of brands should be determined by the individual laboratories based on following consideration: (1) simplicity, (2) throughput, (3) cost, (4) speed, and (5) appropriateness (37). Table 2 lists the currently available PCR-based diagnostic tests and identifies which type of format they are based. The commercial PCR tests described in the next section is not an endorsement of any one product, but rather presents the reader with the commercial kits available for PCR detection of pathogens in foods.

Real-Time PCR. *BAX Dupont-Qualicon* This PCR test is based on the use of pathogen-specific primers combined with a dye that allows for detection of amplicon formation during each cycle. A selective enrichment step appropriate for each food is required before the DNA extraction step. Testing is carried

Table 4.2. Commercial validated and approved test for the detection of bacterial pathogens in food

Test Format	Brand Name	Certification Agency	Organisms Detected
Real-Time PCR	BAX Dupont-Qualicon	AOAC-RI	*Salmonella, E. coli* O157:H7,
		USDA-FSIS AFNOR NorVal	*Listeria monocytogenes*
	Food-proof Roche	AOAC-RI	*Salmonella, E. coli* O157:H7, *Listeria*
	IQ-Check BioRad	AFNOR	*Salmonella*
PCR-ELISA	Probelia BioRad	AFNOR	*Salmonella, E. coli* O157:H7, *Listeria monocytogenes, Campylobacter jejuni* and *Campylobacter coli.*
	AnDiaTec		*Salmonella*
			Listeria monocytogenes

Notes: AOAC-RI: Association of Official Analytical Chemists-Registration International; AFNOR: Association Française de Noramlisation; NorVal: Nordic Validation Organ; USDA-FSIS: United States Department of Agriculture-Animal and Food Safety and Inspection Service.

out in a 96-well type matrix and up to 94 samples can be tested at once with one positive and one negative control. The product amplification is detected real time by including the fluorescent SYBR Green I. This fluorogenic reporter dye is not specific for the desired target molecule, therefore, post-PCR melting curve analysis is required in the protocol, and spurious, nonspecific amplicons are easy to identify. The BAX system has been developed for the detection of *Salmonella* spp., *Listeria monocytogenes*, *E. coli* O157:H7, and *Campylobacter* spp (12, 16, 20) (Table 2).

The BAX *Salmonella* spp has been accepted as an official method by several accreditation bodies; the Association of Official Analytical Chemists (AOAC) has accepted it as an official method (#2003.09) for use in raw beef, raw chicken, raw frozen fish, cheese, frankfurters, and orange juice (5). The AOAC-RI Performance Tested Method license #100201 applies to food: tested on milk, black pepper, chilled ready meal, chipped ham, chocolate, cooked chicken, cooked fish, custard, dry pet food, elbow macaroni, frozen peas, hot dogs, non-fat dry milk, orange juice, peanut butter, pizza dough, seafood-prawns, alfalfa sprouts, ground beef, and liquid egg (3). The USDA-FSIS (MLG 4C.00) has adopted the BAX *Salmonella* spp. for use in ready-to-eat meat, poultry, and pasteurized eggs (6). The AFNOR (certificate QUA-18/3-11/02) applies to all human and animal food (2). The NordVal (certificate 2003-2-5408-00023) applies to all foods and animal feed (7).

The BAX *Listeria monocytogenes* has been approved by the AOAC. It has been accepted as an official method (#2003.12 AOAC-RI Performance Tested Method license #070202) for use in a wide variety of foods including raw meats, fresh produce/vegetables, processed meats, seafood, dairy cultured/noncultured, egg and egg products, and fruit juices (4). The USDA-FSIS (MLG 8A.00) has adopted the BAX *Listeria monocytogenes* for use in red meat, poultry, egg, and environmental samples (11). The BAX *E. coli* O157:H7 has been approved by the AOAC, and has accepted as an official method (#2004.8 AOAC-RI Performance Tested Method license #010402) for use in apple cider, orange juice, and ground beef (9). The USDA-FSIS is currently in the process of validating this technology.

Food-proof Roche This PCR test is based on real-time detection of either *Salmonella* spp. or *Listeria monocytogenes* DNA in raw materials and food samples through the use of a combination of primers and sequence-specific taq-man probes with hot start methodology. An internal control is added to each sample prior to extraction, in order to assess the presence of PCR inhibitors. Additionally, this commercial test contains uracil-DNA glycosidase to avoid PCR carryover contamination. The *Salmonella* spp. test method is certified by the AOAC-RI with license #12030 (8), as a performance-tested method for detecting *Salmonella* in food products. Some raw materials are highly inhibitory for the PCR reaction and the use of a proprietary sample preparation kit (High Pure Food-proof kit; Roche; Indianapolis, IN) seems to ensure DNA of high quality for PCR. The *Listeria monocytogenes* test method has also been certified by the AOAC-RI with license #12030 (10) as a performance-tested method for the detection *Listeria monocytogenes* in food products when used in combination with ShortPrep foodproof II Kit. These foodstuffs include peanut butter, dried whole eggs, dry whole milk, dry pet food, milk chocolate, melon cubes, white cabbage, pizza, vanilla ice cream, paprika emulsion dye, spaghetti, sausage, gravlax, "harzer" cheese, raw ground chicken, raw ground pork, bean sprouts, parsley flakes, ham, and Pollack fillet.

IQ-Check BioRad This commercial test (BioRad; Hercules, CA) uses primers and a molecular beacon probe tagged with a fluorescent label specific for the target organism. Amplified products are detected real time by detection of the fluorescence. This system also contains an internal control, present in the amplification mix that assesses the presence of PCR inhibitors. The internal control is detected real time using another florescent beacon labeled with a different fluoroprobe (27, 45). The BioRad's IQ-Check *Salmonella* detection kit has been approved by AFNOR as a valid method for the detection of *Salmonella* in all human and animal food products, and environmental samples.

PCR-ELISA. *Probelia BioRad* This PCR test is based on the enzymatic detection of a PCR product that combines DNA: DNA hybridization with its capture and in a microtitier plate with an internal oligoprobe. For the detection of *Salmonella*, this PCR-ELISA can detect 3 CFU/25g sample with 99.6% specificity, following an 18 h of preenrichment step (14, 18). It includes an internal control to evaluate PCR inhibitors in samples, which are monitored in a parallel well. Results depend on the optical density obtained on the detection microplate relative to the internal control well. *Salmonella* and *Listeria* applications have been approved AFNOR for all foodstuffs.

AnDiaTec Salmonella sp. PCR-ELISA This commercial kit (AnDiaTec GmbH & Co.; Kornwestheim, Germany) comes in two modules. Module one includes all reagents needed for DNA extraction, amplification mixture in a ready-to-use format, and negative and positive controls. The second module consists of a microtiter plate, probes, the peroxidase conjugate, and all the buffers required for DNA: DNA hybridization and enzymatic detection of the amplified PCR products. For *Salmonella* spp. detection, this test has a demonstrated 98% agreement with bacterial culture when it is conducted according to the ISO 6579 standards. Only samples that had high levels of inhibitors, such as bitter chocolate and herbs required a different extraction method than the one included in the tests kit (35). There is also a kit for the detection of *Listeria monocytogenes*.

REFERENCES

1. Anonymous. 2002. Microbiology of food and animal feeding stuffs—Protocol for the validation of alternative methods (EN ISO/FIDS 16140). European Committee for Standardization, AFNOR, Paris, France.
2. Anonymous. 2002b. AFNOR Certificate QUA-18/3-11/02: Dupont Qualicon BAX® system PCR assay for screening *Salmonella*. www.afnor.fr/portail.asp.
3. Anonymous. 2002c. BAX® system with automated detection PCR assay for screening for *Salmonella*. www.aoac.org/testkits/testedmethods.html.

4. Anonymous. 2002d. BAX® system with automated detection PCR assay for screening for *L. monocytogenes*. www.aoac.org/testkits/testedmethods.html.

5. Anonymous. 2003a. Evaluation of the BAX® system for the detection of *Salmonella* in selected foods (DuPont Qualicon, Inc.; Willmington, DE). www.aoac.org/vmeth/ newsmtd.htm#faoma.

6. Anonymous. 2003b. FSIS procedure for the use of the BAX® system PCR assay for screening *Salmonella* in raw meat, carcass sponge samples, whole bird rinses, ready-to-eat meat and poultry products and pasteurized egg products. www.fsis.usda.gov/ ophs/microlab/mlg4c01.pdf.

7. Anonymous. 2003c. NordVal Certificate 2003-20-5408-00023: Dupont Qualicon BAX® system PCR assay for screening *Salmonella*. www.nmkl.org/NordVal/ NordVal.htm.

8. Anonymous. 2003d. Roche Diagnostics LightCycler® foodproof *Salmonella* detection kit for *Salmonella spp.* in combination with ShortPrep foodproof I Kit. www.aoac.org/testkits/testedmethods.html

9. Anonymous. 2004a. BAX® system PCR assay for screening *E. coli* O157:H7 kit. www.aoac.org/testkits/testedmethods.html.

10. Anonymous. 2004b. Roche/BIOTECON diagnostics LightCycler® *Listeria monocytogenes* detection systems in combination with the *Listeria* ShortPrep foodproof® II kit. www.aoac.org/testkits/testedmethods.html.

11. Anonymous. 2005. FSIS procedure for the use of *Listeria monocytogenes* BAX® screening test. www.fsis.usda.gov/Ophs/Microlab/Mlg_8A_01.pdf.

12. Bhagwat, A.A. 2003. Simultaneous detection of *Escherichia coli* O157:H7, *Listeria monocytogenes* and *Salmonella* strains by real-time PCR. Int. J. Food Microbiol. **84**:217–224.

13. Borst, A., A.T. Box, and A.C. Fluit. 2004. False-positive results and contamination in nucleic-acid amplification assays: Suggestions for a "prevent and destroy" strategy. Eur. J. Clin. Microbiol. Infect. Dis. **23**:289–299.

14. Coquard, D., A. Exinger, and J.M. Jeltsch. 1999. Routine detection of *Salmonella* species in water: Comparative evaluation of the ISO and PROBELIA polymerase chain reaction methods. J. AOAC Int. **82**:871–876.

15. Cui, S., C.M. Schroeder, D.Y. Zhang, and J. Meng. 2003. Rapid sample preparation method for PCR-based detection of *Escherichia coli* O157:H7 in ground beef. J. Appl. Microbiol. **95**:129–134.

16. Englen, M.D. and P.J. Fedorka-Cray. 2002. Evaluation of commercial diagnostic PCR for the identification of *Campylobacter jejuni* and *Campylobacter coli*. Lett. Appl. Microbiol. **55**:353–356.

17. Erlandsson, A., A. Backman, E. Tornqvist, and P. Olsen. 1997. PCR assay or culture for diagnosis of *Bordetella pertussis* in the routine diagnostic laboratory? J. Infect. **35**:221–224.

18. Fach, P., F. Dilasser, J. Grout, and J. Tache. 1999. Evaluation of a polymerase chain reaction-based test for detecting *Salmonella spp.* in food samples: Probleia *Salmonella spp.* J. Food Prot. **62**:1387–1393.

19. Herrera-Leon, S., J.R. McQuiston, M.A. Usera, P.I. Fields, J. Garaizar, and M.A. Echeita. 2004. Multiplex PCR for distinguishing the most common phase-1 flagellar antigens of *Salmonella spp.* J. Clin. Microbiol. **42**:2581–2586.

20. Hochberg, A.M., A. Roering, V. Gangar, M. Curiale, and W.M. Barbour. 2001. Sensitivity and specificity of the BAX for screening/*Listeria monocytogenes* assay: Internal validation and independent laboratory study. J. AOAC Int. **84**:1087–1097.

21. Hong, Y., T. Liu, C. Hofacre, M. Maier, S. Ayers, D.G. White, L. Wang, and J. J. Maurer. 2003. A restriction fragment length polymorphism based polymerase chain reaction as an alternative to serotyping for identifying *Salmonella* serotypes. Avian Dis. **47**:387–395.

22. Hoorfar, J., N. Cook, B. Malorny, M. Wagner, M.D. De, A. Abdulmawjood, and P. Fach. 2003. Making internal amplification control mandatory for diagnostic PCR. J. Clin. Microbiol. **41**:5835.

23. Hoorfar, J., N. Cook, B. Malorny, M. Wagner, M.D. De, A. Abdulmawjood, and P. Fach. 2004. Diagnostic PCR: Making internal amplification control mandatory. Lett. Appl. Microbiol. **38**:79–80.

24. Hoorfar, J. and N. Cook. 2003. Critical aspects of standardization of PCR. *In* K. Sachese and J. Frey (eds.), PCR Detection of Microbial Pathogens: Methods in Molecular Microbiology, Vol. 216, pp. 51–64.

25. Josefsen, M.H., S.T. Lambertz, S. Jensen, and J. Hoorfar. 2003. Food-PCR. Validation and standardization of diagnostic PCR for detection of Yersinia enterocolitica and other foodborne pathogens. Adv. Exp. Med. Biol. **529**:443–449.

26. Keramas, G., D.D. Bang, M. Lund, M. Madsen, H. Bunkenborg, P. Telleman, and C.B. Christensen. 2004. Use of culture, PCR analysis, and DNA microarrays for detection of *Campylobacter jejuni* and *Campylobacter coli* from chicken feces. J. Clin. Microbiol. **42**:3985–3991.

27. Liming, S.H. and A.A. Bhagwat. 2004. Application of a molecular beacon-real-time PCR technology to detect *Salmonella* species contaminating fruits and vegetables. Int. J. Food Microbiol. **95**:177–187.

28. Liu, T., K. Liljebjelke, E. Bartlett, C.L. Hofacre, S. Sanchez, and J.J. Maurer. 2002. Application of nested PCR to detection of *Salmonella* in poultry environments. J. Food Prot. **65**:1227–1232.

29. Lo, Y.M. 1998. Methods in Molecular Medicine: Clinical Applications of PCR, Vol. 16. Humana Press Inc., Totowa, NJ.

30. Loeffler, J., K. Schmidt, H. Hebart, U. Schumacher, and H. Einsele. 2002. Automated extraction of genomic DNA from medically important yeast species and filamentous fungi by using the MagNA pure LC system. J. Clin. Microbiol. **40**:2240–2243.

31. Luk, J.M., U. Kongmuang, R.S. Tsang, and A.A. Lindberg. 1997. An enzyme-linked immunosorbent assay to detect PCR products of the *rfbS* gene from serogroup D salmonella: A rapid screening prototype. J. Clin. Microbiol. **35**:714–718.

32. Macrina, F.L. 1995. Scientific Integrity: An Introductory Text with Cases. ASM Press, Washington, DC.

33. Malorny, B., E. Paccassoni, P. Fach, C. Bunge, A. Martin, and R. Helmuth. 2004. Diagnostic real-time PCR for detection of *Salmonella* in food. Appl. Environ. Microbiol. **70**:7046–7052.

34. Malorny, B., P.T. Tassios, P. Radstrom, N. Cook, M. Wagner, and J. Hoorfar. 2003. Standardization of diagnostic PCR for the detection of foodborne pathogens. Int. J. Food.

35. Metzger-Boddien, C., A. Bostel, and J. Kehle. 2004. AnDiaTec *Salmonella sp.* PCR-ELISA for analysis of food samples. J. Food Prot. **67**:1585–1590.

36. Nogva, H.K., K. Rudi, K. Naterstad, A. Holck, and D. Lillehaug. 2000. Application of 5'-nuclease PCR for quantitative detection of *Listeria monocytogenes* in pure cultures, water, skim milk, and unpasteurized whole milk. Appl. Enivorn. Microbiol. **66**:4266–4271.

37. Riley, L. W. 2004. Molecular epidemiology of infectious diseases: principles and practices. ASM Press, Washington, DC.

38. Rudi, K., H.K. Hoidal, T. Katla, B.K. Johansen, J. Nordal, and K.S. Jakobsen. 2004. Direct real-time PCR quantification of *Campylobacter jejuni* in chicken fecal and cecal samples by integrated cell concentration and DNA purification. Appl. Environ. Microbiol. **70**:790–797.

39. Sails, A.D., A.J. Fox, F.J. Bolton, D.R. Wareing, and D.L. Greenway. 2003. A real-time PCR assay for the detection of *Campylobacter jejuni* in foods after enrichment culture. Appl. Environ. Microbiol. **69**:1383–1390.

40. Sarkar, G. and S.S. Sommer. 1993. Removal of DNA contamination in polymerase chain reaction reagents by ultraviolet irradiation. Methods Enzymol. **218**:381–388.

41. Sergeev, N., M. Distler, S. Courtney, S.F. Al-Khaldi, D. Volokhov, V. Chizhikov, and A. Rasooly. 2004. Multipathogen oligonucleotide microarray for environmental and biodefense applications. Biosens. Bioelectron. **20**:684–698.

42. Shearer, A.E., C.M. Strapp, and R.D. Joerger. 2001. Evaluation of a polymerase chain reaction-based system for detection of *Salmonella enteritidis*, *Escherichia coli* O157:H7, *Listeria spp.*, and *Listeria monocytogenes* on fresh fruits and vegetables. J. Food Prot. **64**:788–795.

43. Shi, P.Y., E.B. Kauffman, P. Ren, A. Felton, J.H. Tai, A.P. Dupuis, II, S.A. Jones, K.A. Ngo, D.C. Nicholas, J. Maffei, G.D. Ebel, K.A. Bernard, and L.D. Kramer. 2001. High-throughput detection of West Nile virus RNA. J. Clin. Microbiol. **39**:1264–1271.

44. Silbernagel, K., R. Jechorek, C. Carver, W.M. Barbour, and P. Mrozinski. 2003. Evaluation of the BAX system for detection of *Salmonella* in selected foods: Collaborative study. J. AOAC Int. **86**:1149–1159.

45. Uyttendaele, M., K. Vanwildemeersch, and J. Debevere. 2003. Evaluation of real-time PCR vs. automated ELISA and a conventional culture method using a semi-solid medium for detection of *Salmonella*. Lett. Appl. Microbiol. **37**:386–391.

46. Wellinghausen, N., B. Wirths, A. Essig, and L. Wassill. 2004. Evaluation of the Hyplex Bloodscreen multiplex PCR-enzyme-linked immunosorbent assay system for direct identification of gram-positive cocci and gram-negative bacilli from positive blood cultures. J. Clin. Microbiol. **42**:3147–3152.

47. Wiedbrauk, D.L. and R.L. Hodinka. 1998. Applications of the polymerase chain reaction. *In* S. Specter, M. Bendinelli, and H. Friedman (eds.), Rapid Detection of Infectious Agents. Plenum Press, New York.

48. Wilson, I.G. 1997. Inhibition and facilitation of nucleic acid amplification. Appl. Environ. Microbiol. **63**:3741–3751.

49. Wolk, D.M., S.K. Schneider, N.L. Wengenack, L.M. Sloan, and J.E. Rosenblatt. 2002. Real-time PCR method for detection of *Encephalitozoon intestinalis* from stool specimens. J. Clin. Microbiol. **40**:3922–39288.

CHAPTER 5

Molecular Detection of Foodborne Bacterial Pathogens

Azlin Mustapha*, Ph.D. and Yong Li, Ph.D.
Food Science Program, University of Missouri-Columbia,[1] Columbia, MO 65211
[1]*Food Science and Human Nutrition Program, University of Manoa, Honolulu, HI 96822*

Introduction
Conventional PCR Detection of Foodborne Pathogens
Multiplex PCR Detection of Foodborne Pathogens
Reverse Transcriptase PCR Detection of Foodborne Pathogens
Real-Time PCR Detection of Foodborne Pathogens
Conclusions
References

INTRODUCTION

Since its discovery in the mid 1980s, the Nobel prize-winning polymerase chain reaction (PCR) has gained acceptance as a powerful microbial detection tool. The application of automated PCR technology in the medical and pharmaceutical industries has a longer history than its use in the food industry. In recent years, however, PCR technology has become more recognized for its potential at becoming a powerful alternative to cultural methods of pathogen detection in foods. The BAX PCR system manufactured by Dupont Qualicon, Inc. and the TaqMan Pathogen Detection kits by Applied Biosystems Co. are two examples of automated PCR systems that have found application in the food industry. The BAX PCR is currently an AOAC-approved PCR-based method that can be used to detect *Listeria monocytogenes*, *Listeria* species, *Salmonella* and *Escherichia coli* O157:H7 in food products, and is used by the USDA Food Safety and Inspection Service (FSIS) for detecting *L. monocytogenes* and *Salmonella* in meat, poultry, egg, and ready-to-eat meat products.

The advantage of using PCR techniques for food products is the specificity and rapidity of the tests as compared to traditional cultural techniques. However, the sensitivity of a PCR-based test for detection of pathogens in foods will depend on the type of food matrix involved. Because of the complexity of food matrices, many compounds in foods can prove to be inhibitory to PCR reactions (74). Thus, in most PCR assays, including the automated BAX and

---

*Corresponding author. Food Science Program, 256 William Stringer Wing, University of Missouri-Columbia, Columbia, MO 65211, Phone: (573) 882-2649; FAX: (573) 882-0596; e-mail: MustaphaA@missouri.edu.

TaqMan systems, food samples are enriched prior to the amplification steps in order to overcome this hurdle. In addition to increasing the number of target microorganisms, and thus the detection sensitivity, enrichment is also helpful at reducing the risk of amplifying nucleic acids from dead or nonculturable cells (11). Following PCR, target species can be detected by agarose gel electrophoresis or hybridizations with labeled DNA probes (60).

Despite its limited use in the food industry, numerous studies have been reported in the literature on the development of PCR-based detection methods for foodborne pathogens. The following text will discuss the various PCR detection methods that have been successful at detecting various pathogens in different types of foods. Table 1 illustrates the target genes, primer sequences, tested foods, detection limits, specificity, and references of certain published PCR protocols.

CONVENTIONAL PCR DETECTION OF FOODBORNE PATHOGENS

Conventional PCR relies on the amplification of nucleic acids via a single pair of primers to detect one pathogen at a time, and the PCR reaction is optimized for the specific food product tested. With enrichment of tested samples, conventional PCR assays can detect *Clostridium perfringens* at 10 cfu/g in meat, milk, and salad (17), enterohemorrhagic *E. coli* O157:H7 at *ca* 10^{-1} cfu/g in beef (55), enterotoxigenic *E. coli* at 1 cfu/ml in milk (70), *L. monocytogenes* at *ca* 10^{-2} to 10^0 cfu/g in various foods (19, 30, 48, 63), *Salmonella* at *ca* 10^{-1} to 10^1 cfu/g in milk and meat products (11, 22, 43, 44), *Shigella* at 2 cfu/g in mayonnaise (71), *Staphylococcus aureus* at 5–15 cfu/g in skim milk and cream (65), and 10 cfu/ml in raw milk and curd (16), and *Vibrio parahaemolyticus* at 10 cfu in fish (31). However, other studies have found much higher detection limits of PCR assays for foods, without enrichment, including 4×10^2 to 4×10^3 cfu/g for *C. perfringens* in Korean ethnic foods (32), 10 cfu/ml for *L. monocytogenes* in milk (3), 5×10^1 to 5×10^2 cfu/ml for *Shigella* in various produce washes (36), 10^2 cfu/g for *S. aureus* in skim milk and cheddar cheese (67), 3×10^2 cfu/g for *V. parahaemolyticus* in shellfish (74), and 4×10^4 cfu/g for *Yersinia enterocolitica* in pork (37). A PCR based on degenerate primers targeting known nonribosomal peptide synthetases (NRPS) has also been successfully developed for detecting emetic strains of *Bacillus cereus* (15).

MULTIPLEX PCR DETECTION OF FOODBORNE PATHOGENS

In multiplex PCR, two or more gene loci are simultaneously amplified in one reaction. This technique has been used widely to characterize pathogenic bacteria on the basis of their virulence factors and antigenic traits. Fratamico et al. (21) employed primers for a plasmid-encoded hemolysin gene ($hlyA_{933}$),

Table 5.1. Representative PCR methods for common foodborne bacterial pathogens

Bacterial Pathogen	Target Gene	Primer sequence (5' to 3')	Tested Food	Detection Limit[1]	Specificity	Reference
Bacillus cereus	cerAB	F: GAGTTAGAGAACGGTATTTATGCTGC R: GCATCCCAAGTCGCTGTATGTCCAG	milk	1 cfu/ml[2,3]	B. cereus group	60
	16S rRNA	F: TCGAAATTGAAAGGCGGC R: GGTGCCAGCTTATTCAAC	NA[4]	NA	B. cereus group	24
	groEL	F: TGCAACTGTATTAGCACAAGCT R: TACCACGAAGTTTGTTCACTACT	NA	NA	B. cereus group	9
	gyrB	F: GTTTCTGCTGGTTTACATGG R: TTTTGAGCGATTTAAATGC	coffee concentrate	10 cfu/ml[2]	B. cereus group	51
	NA	F: GACAAGAGAAATTTCTACGAGCAAGT R: GCAGCCTTCCAATTACTCCTTCTGCCACAGT	NA	NA	Emetic toxin producing B. cereus	15
Campylobacter	flaA/B	F: CCAAATCGGTTCAAGTTCAAATCAAAC R: CCACTACCTACTGAAAATCCCGAACC	NA	5–20 cells[3]	C. jejuni, C. coli	58
	16S rRNA	F1: AATCTAATGGCTTAACCATTA R1: GTAACTAGTTTAGTATTCCGG	NA	NA	C. jejuni, C. coli	46
	hippurase	F2: GAAGAGGTTTGGGTGGTG R2: AGCTAGCTTCGCATAATAACTTG				
	ask	F3: GGTATGATTTCTACAAAGCGAG R3: ATAAAAGACTATCGTCGCGTG				
	16S rRNA	F1: ATCTAATGGCTTAACCATTAAAC R1: GGACGGTAACTAGTTTAGTATT	NA	NA	C. jejuni, C. coli	14

Continued

Table 5.1. Representative PCR methods for common foodborne bacterial pathogens—cont'd

Bacterial Pathogen	Target Gene	Primer sequence (5' to 3')	Tested Food	Detection Limit[1]	Specificity	Reference
	mapA	F2: CTATTTATTTTGAGTGCTTGTG R2: GCTTTATTTGCCATTTGTTTTATTA				
	ceuE	F3: AATTGAAAATTGCTCCAACTATG R3: TGATTTTATTATTTGTAGCAGCG				
	ccoN	F: TTGGTATGGCTATAGGAACTCTTATAGCT R: CACACCTGAAGTATGAAGTGGTCTAAGT PROBE: TGGCCATATCCTAATTTAAATTATTTACCAGGAC	Raw chicken, offal, shellfish, raw meat, milk	NA	C. jejuni	59
	omp50	F: TGTAAAAGCTGAACTTGCC R: GCCGTTCCTCTTGTCATTC	NA	NA	Campylobacter	5, 13
	ccoN	F: AGAACACGCGGACCTATATA R: CGATGCATCCAGGAATGTAT	water, milk	25 to 2 × 10³ cfu²	C. jejuni, C. coli, C. upsalensis	27
Clostridium perfringens	etxB/D	F: TACTCATACTGTGGGAACTTCGATACAAGC R: CTCATCTCCCATAACTGCACTATAATTTCC	NA	NA	Type B or D strains of C. perfringens	23
	pls	F1: AAGTTACCTTTGCTGCATAATCCC	cooked food, pork butchery, raw meat, milk, salad	10 cfu/g	C. perfringens	17

Organism	Gene	Primer	Sequence	Food	Detection limit	Target	Ref.
	cpe	R1:	ATAGATACTCCATATCATCCTGCT				
		F2:	GAAAGATCTGTATCTACAACTGCTGGTCC				
		R2:	GCTGGCTAAGATTCTATATTTTTGTCCAGT				
	cpa	F1:	TGCTAATGTTACTGCCGTTGATAG	NA	NA	C. perfringens producing different toxins	70
	cpb	R1:	ATAATCCCAATCATCCCAACTATG				
		F2:	AGGAGGTTTTTTATGAAG				
		R2:	TCTAAATAGCTGTTACTTTGT				
	etx	F3:	TACTCATACTGTGGGAACTTCGATACAAGC				
		R3:	CTCATCTCCCATAACTGCACTATAATTTCC				
	iap	F4:	TTTTAACTAGTTCATTTCCTAGTTA				
		R4:	TTTTTGTATTCTTTTTCTCTAGATT				
	cpe	F:	ACTTAGAGTATCTATAAACTTGATACTC	Korean ethnic foods	4×10^2 cfu/g to 4.5×10^3	Enterotoxigenic C. perfringens cfu/g[2]	32
		R:	TAAATTGTTACTAAGCATATTATAATTAACATC				
Pathogenic Escherichia coli	malB	F1:	TGACCACACGCTGACGCTGACCA	ground beef	10–100 cfu/g	total E. coli, ETEC, EIEC, EHEC	40
	hlt	R1:	TTACATGACCTCGGTTTAGTTCACAGA				
		F2:	TTACGGCGTTACTATCCTCTA				
		R2:	GGTCTCGGTCAGATATGTGATTC				
	invx	F3:	TCCTGCTTAGATGATGGAGTAAT				
		R3:	CTCACCATACCATCCAGAAAGAAG				
	vt	F4:	TTAACCACCCACGGCAGT				
		R4:	GCTCTGGATGCATCTCTGGT				

Continued

Table 5.1. Representative PCR methods for common foodborne bacterial pathogens—cont'd

Bacterial Pathogen	Target Gene	Primer sequence (5' to 3')	Tested Food	Detection Limit[1]	Specificity	Reference
Shiga-toxin-producing E. coli	stx-2	F1: TGTTTATGGCGGTTTTATTTG	ground beef	1 cfu/g[3,6]	viable Shiga-toxin-producing E. coli	52
		R1: ATTATTAAACTGCACTTCAG				
	stx-1/2	F2: GGATCCTTTACGATAGACCTTCTCGAC				
		R2: GGATCCCACATATAAATTATTTCGCTC				
Entero-hemorrhagic E. coli (EHEC)	stx 1	F1: GACTGCAAAGACGTATGTAGATTCG	ground beef	1–10 cfu/g[5]	EHEC O157, O111, O26	62
		R1: ATCTATCCCTCTGACATCCAACTGC				
	stx 2	F2: ATTAACCACACCCCACCG				
		R2: GTCATGGAAACCGTTGTCAC				
	eaeA_O26	F3: CTCTGCCAAAGAACTGGTTACAG				
		R3: TTTCCATGTGTATTTTCCATTGC				
	eaeA_O111	F4: GCTCCGAATTATATGATAAGAGTGG				
		R4: TCTGTGAGGATGGTAATAAATTTCC				
	eaeA_O157	F5: GTAAGTTACACTATAAAAGCACCGTCG				
		R5: TCTGTGTGGATGGTAATAAATTTTTG				
EHEC O157:H7	stx 1	F1: CAGTTAATGTGGTGGCGAAGG	NA	NA	EHEC O157: H7, O157:NM	8
		R1: CACCAGACAATGTAACCGCTG				
	stx 2	F2: ATCCTATTCCCGGGAGTTTACG				
		R2: GCGTCATCGTATACACAGGAGC				
	uidA	F3: GCGAAAACTGTGGAATTGGG				
		R3: TGATGCTCCATAACTTCCTG				

Organism	Gene	Primer	Sequence	Food	Detection limit	Pathogen	Ref
	eaeA	F1:	CCATAATCATTTTATTTAGAGGGA	NA	NA	EHEC O157: H7, O157:NM	54
		R1:	GAGAAATAAATTATATTAATAGATCGGA				
	stx 1	F2:	TGTAACTGGAAAGGTGGAGTATACA				
		R2:	GCTATTCTGAGTCAACGAAAAATAAC				
	stx 2	F3:	GTTTTTCTTCGGTATCCTATTCC				
		R3:	GATGCATCTCTGGTCATTGTATTAC				
	fliC$_{h7}$	F1:	GCGCTGTCGAGTTCTATCGAGC	ground beef, blue cheese, mussels, alfalfa sprouts,	1 cfu/g	EHEC O157:H7	21
		R1:	CAACGGTGACTTTATCGCCATTCC				
	stx 1	F2:	TGTAACTGGAAAGGTGGAGTATACA				
		R2:	GCTATTCTGAGTCAACGAAAAATAAC				
	stx 2	F3:	GTTTTTCTTCGGTATCCTATTCC				
		R3:	GATGCATCTCTGGTCATTGTATTAC				
	eaeA	F4:	ATTACCATCCACACAGACGGT				
		R4:	ACAGCGTGGTTGGATCAACCT				
	hlyA$_{O157}$	F5:	ACGATGTGGTTTATTCTGGA				
		R5:	CTTCACGTCACCATACATAT				
	NA	F:	AGCACTGAATGACGCGCAATTGAGACA	beef	10^{-1} cfu/g	EHEC O157: H7, O157:NM	55
		R:	TCTGAGGGACCTTAATTTTCCCTGATTCTC				
Enterotoxigenic E. coli (ETEC)	elt I	F1:	GCTGACTCTAGACCCCCAG	milk	1 cfu/ml	ETEC	69
		R1:	TGTAACCATCCTCTGCCGGA				

Continued

Table 5.1. Representative PCR methods for common foodborne bacterial pathogens—cont'd

Bacterial Pathogen	Target Gene	Primer sequence (5' to 3')	Tested Food	Detection Limit[1]	Specificity	Reference
	est II	F2: CTGTGTGAACATTATAGACAAATA R2: ACCATTATTTGGGCGCCAAAG	NA	NA	ETEC	56
	uspA	F1: CCGATACGCTGCCAATCAGT R1: ACGCAGACCGTAAGGGCCAGAT				
	elt I	F2: TATCCTCTCTATATGCACAG R2: CTGTAGTGGAAGCTGTTATA				
	est I	F3: TCTTTCCCCTCTTTAGTCAG R3: ACAGGCCGGATTACAACAAAG				
	est II	F4: GCCTATGCATCTACACAATC R4: TGAGAAATGGACAATGTCCG				
Listeria monocytogenes	iap	F: CAAACTGCTAACACAGCTACT R: GCACTTGAATTGCTGTTATTG	NA	NA	L. monocytogenes	6
	hlyA	F: CTAATCAAGACAATAAAATC R: GTTAGTTCTACATCACCTGA	cheese	1.6×10^0 cfu/g	L. monocytogenes	19
	iap	F: CAAACTGCTAACACAGCTACT R: GCACTTGAATTGCTGTTATTG	cooked ground beef	3 cfu/g	viable L. monocytogenes	34
	hlyA	F: GGGAAATCTGTCTCAGGTGATGT R: CGATGATTTGAACTTCATCTTTTGC	cabbage	6 cfu/g[2,5]	L. monocytogenes	25
	inlAB	F: CTTCAGGCGGATAGATTAGG R: TTCGCAAGTGAGCTTACGTC	frank-furters	4×10^{-1} cfu/g	L. monocytogenes	30

	Gene	Primer		Food	Detection limit	Pathogen	Ref
	16S rRNA	F1:	GCTAATACCGAATGATAAGA	fresh and ready-to-eat meat and fish, potato salads, vegetable salads, pasta, ice cream	4×10^{-2} to 2×10^{-1} cfu/g	L. monocytogenes	63
		F2:	GGCTAATACCGAATGATGAA				
		R:	AAGCAGTTACTCTTATCCT				
	actA	F:	GTGATAAAATCGACGAAAATCC	soft cheeses	4×10^{-2} to 4 cfu/g	L. monocytogenes	48
		R:	CTTGTAAAACTAGAAATCTAGCG				
	hlyA	F:	TTGCCAGGAATGACTAATCAAG	milk	10 cfu/ml[2,6]	L. monocytogenes	3
		R:	ATTCACTGTAAGCCATTTCGTC				
	transcriptional regulatory gene	F:	CGCAAGAAGAAATTGCCATC	NA	10 pg DNA	L. monocytogenes	47
		R:	TCCGCGTTAGAAAAATTCCA				
Salmonella	NA	F:	AGCCAACCATTGCTAAATTGGCGCA	beef, pork	NA	Salmonella	1, 2
		R:	GGTAGAAATTCCCAGCGGGTACTG				
	oriC	F:	TTATTAGGATCGCGCCAGGC	chicken	10^{-1} cfu/g[6]	Salmonella	20
		R:	AAAGAATAACCGTTGTTCAC				
	repeat sequence	F:	GATCATCCATTCGGCATTAAACA	frozen chicken	3 cfu/g	Salmonella	28
		R:	CTTCAGCGACGGAAGGGTAAATC				

Continued

Table 5.1. Representative PCR methods for common foodborne bacterial pathogens—*cont'd*

Bacterial Pathogen	Target Gene	Primer sequence (5' to 3')	Tested Food	Detection Limit[1]	Specificity	Reference
	ompC	F: ACCGCTAACGCTCGCCTGTAT R: AGAGGTGGACGGGTTGCTGCCGTT	ground beef	20 cfu/g	*Salmonella*	35
	invA	F: GCTGCGCGCGAACGGCCAAG R: TCCCGGCAGAGTTCCCATT	pork, beef, poultry meat, fermented sausage, fish	4×10^{-1} cfu/g	*Salmonella*	11
	invA	F: CGGTGGTTTTAAGCGTACTCTT R: CGAATATGCTCCACAAGGTTA	ground beef, apple cider	1 cfu/g	*Salmonella*	21
	16S rRNA	F: GTGTTGTGGTTAATAACCGCAGCA R: TGTTBGMTCCCCACGCTTTCG	whole milk, chicken	1 to 9 cfu/g	*Salmonella*	43
	rfbS	F: TCACGACTTACATCCTAC R: CTGCTATATCAGCACAAC	NA	10 cfu[3]	*Salmonella* serotype D	49
	sefA	F: GGCTTCGGTATCTGGTGTGTA R: GGTCATTAATATTGGCCCTGAATA	egg	1.7×10^{-3} cfu/g[5]	*Salmonella* serotype D	61
	orf6e	F: GCCGTACACGACCTTATAGA R: ACCTACAGGGCACAATAAC	NA	NA	*Salmonella* Enteritidis	64
	fliC	F: CGGTGTTGCCCAGGTTGGTAAT R: ACTGGTAAAGATGGCT	NA	NA	*Salmonella* Typhimurium	64

		Primer sequences	Food	Detection limit	Target	Ref
	mdh	F: TGCCAACGGAAGTTGAAGTG R: CGCATTCCACCACGCCCTTC	milk, chicken meat	10 cfu/g	*Salmonella* Typhimurium	44
Shigella	*spa*	F: AGCGATCTTACGTCTTG R: CGAGATGTGGAGGCAT	carrot, celery, cauliflower, radish, broccoli, coleslaw	NA	*Shigella*	18
	rfc	F1: ATCAGGTGTCGTAATTTTA R1: GGGCTAAGTTCCCTC F2: ATTGGTGGTGGTGGAAGATTACTGG R2: TTTTGCTCCAGAAGTGAGG F3: AGCTAATGCGTTTTGGGGAAT R3: TCCCAATGACTGATACCATGG	NA	$<10^4$ cfu/g^2	*Shigella*	26
	virA	F: CTGCATTCTGGCAATCTCTTCACATC R: TGATGAGCTAACTTCGTAAGCCCTCC	mayonnaise	2 cfu/g	*Shigella* and EIEC	71
	ial	F1: CTGGTAGGTATGGTGAGG R1: CCAGGCCAACAATTATTTCC F2: TTTTTAATTAAGAGTGGGGTTTGA R2: GAACCTATGTCTACCTTACCAGAAGT	lettuce, shrimps, milk, blue cheese	10 cfu/g^7	*Shigella* and EIEC	45

Continued

Table 5.1. Representative PCR methods for common foodborne bacterial pathogens—cont'd

Bacterial Pathogen	Target Gene	Primer sequence (5' to 3')	Tested Food	Detection Limit[1]	Specificity	Reference
	ipaH	F: GTTCCTTGACCGCCTTTCCGATACCGTC	cilantro, tomato, beef, lettuce, alfalfa sprouts, apple cider, bean sprouts	50–500 cfu/ml[2]	Shigella and EIEC	36
		R: GCCGGTCAGCCACCCTCTGAGAGTAC				
Staphylococcus aureus	sea	F1: CCTTTGGAAACGGTTAAAACG	NA	100 pg DNA[3]	S. aureus carrying enterotoxin A to E genes	4
		R1: TCTGAACCTTCCCATCAAAAAC				
	seb	F2: TCGCCATCAAACTGACAAACG				
		R2: GCAGGTACTCTATAAGTGCCTGC				
	sec	F3: CTCAAGAACTAGACATAAAAGCTAGG				
		R3: TCAAAATCGGATTAACATTATCC				
	sed	F4: CTAGTTTGGTAATATCTCCTTTAAACG				
		R4: TTAATGCTATATCTTATAGGGTAAACATC				
	see	F5: CAGTACCTATAGATAAGTTAAAACAAGC				
		R5: TAACTTACCGTGGACCCTTC				
	tst	F1: AAGCCCTTTGTTGCTTGCG	NA	100 pg DNA[3]	S. aureus carrying exfoliative toxin A and B genes and the toxic shock syndrome toxin 1 gene	4

Gene	Primers	Food matrix	Detection limit	Organism	Ref.
eta	R1: ATCGAACTTGGCCCATACTT F2: CTAGTGCATTTGTTATTCAAGACG R2: TGCATTGACACCATAGTACTTATTC				
etb	F3: ACGGCTATATACATTCAATTCAATG R3: AAAGTTATTCATTTAATGCACTGTCTC				
23S rRNA	F1: ACGGAGTTACAAAGGACGAC R1: AGCTCAGCCTTAACGAGTAC F2: ATCATCTGGAAAGATGAATCAA R2: ATCGATTAAAACGATTATAGGT	skim milk, cream	5–15 cfu/g[7]	S. aureus	65
entC	F1: ACACCCAACGTATTAGCAGAGAGC R1: CCTGGTGCAGGCATCATATCATAC F2: AGTATATAGTGCAACTTCAACTAA R2: ATCAGCGTTGTCTTCGCTCCAAAT	skim milk, cheddar cheese	100 cfu/g[2]	enterotoxigenic S. aureus	67
nuc					
seg	F1: GCTATCGACACACTACAACC	NA	10[2] cfu/ml for pure culture	S. aureus carrying enterotoxin G to I genes	10
seh	R1: CCAAGTGATTGTCTATTGTCG F2: CACATCATATGCGAAAGC R2: CGAATGAGTAATCTCTAGG				
sei	F3: GATACTGGAACAGGACAAGC R3: CTTACAGGCAGTCCATCTCC				
tdh	F: TTTCATGATTATTCAGTTT R: TTTGTTGGATATACACAT	fish	10 cfu	V. parahaemolyticus	31
pathogenic Vibrio species					

Continued

Table 5.1. Representative PCR methods for common foodborne bacterial pathogens—cont'd

Bacterial Pathogen	Target Gene	Primer sequence (5' to 3')	Tested Food	Detection Limit[1]	Specificity	Reference
	vvhA	F1: GACTATCGCATCAACAACCG R1: AGGTAGCGAGTATTACTGCC F2: GCTATTTCACCGCCGCTCAC R2: CCGCAGAGCCGTAAACCGAA	seafood	10^1–10^2 cfu/g[2,7]	V. vulnificus	39
	toxR	F: GTCTTCTGACGCAATCGTTG R: ATACGAGTGGTTGCTGTCATG	NA	NA	V. para-haemolyticus	33
	hlyA	F: TGCGTTAAACACGAAGCGAT R: AAGTCTTACATTGTGCTTGGGTCA	oyster	6–8 cfu/g[2,5]	V. cholerae	50
	vvhA	F: TGTTTATGTGAGAACGGTGACA R: TTCTTTATCTAGGCCCCAAACTTG	oyster	10^3 cfu/g[2,5]	V. vulnificus	7
	ORF8	F: GTTCGCATACAGTTGAGG R: AAGTACAGCAGGAGTGAG	shellfish	3×10^2 cfu/g[2]	V. para-haemolyticus O3:K6	74
	elastase gene	F: AAACTCAGGTCTGATATACAGC R: AAGTTGCTACCTGGCGTGTTG	NA	25 cfu for pure culture	V. vulnificus	38
	tdh trh tlh	F1: CATCTTCGTACGGTTTCTTTTTACA R1: TCTGTCCCTTTTCCTGCCC F2: GCCAAGTGTAACGTATTTGGATGA R2: TGCCCATTTCCGCTCTCA F3: CGAGAACGCAGACATTACGTTC R3: TGCTCCAGATCGTGTGGTTG	mussel	NA	V. para-haemolyticus	12

	vvh	F: TTCCAACTTCAAACCGAACTATGA R: ATTCCAGTCGATGCGAATACGTTG	oyster	1 cfu/g[5]	*V. vulnificus*	57
Yersinia enterocolitica	16S *rRNA*	F1: GGAATTTAGCAGAGATGCTTTA R1: GGACTACGACAGACTTTATCT	pork	4×10^4 cfu/g[2]	*Y. enterocolitica*	37
	yadA	F2: TGTTCTCATCTCCATATGCATT R2: TTCTTTCTTTAATTGCGCGACA				
	ail	F: GGTCATGGTGATGTTGATTACTATTCA R: CGGCCCCCAGTAATACCATA	ground pork	≤1 cfu/g[5]	*Y. enterocolitica*	29
	yst	F: AATGCTGTCTTCATTTGGAGC R: ATCCCAATCACTACTGACTTC	ground pork, tofu	10^3 cfu/g[2, 5]	*Y. enterocolitica*	72

[1] Samples were enriched prior to the PCR assay to determine the detection limit, unless otherwise specified.
[2] Samples were not enriched prior to the PCR assay to determine the detection limit.
[3] Hybridization was performed to identify the PCR products.
[4] Not available.
[5] In the format of fluorogenic real-time PCR.
[6] Samples were processed via immunomagnetic separation.
[7] In the format of nested PCR

chromosomal flagella ($fliC_{H7}$; flagellar structural gene of H7 serotype), Shiga toxins (stx_1, stx_2), and attaching and effacing ($eaeA$) genes for specific identification of *E. coli* O157:H7. Similar protocols were reported for concurrent determination of multiple toxin genes in *C. perfringens* (70) and *S. aureus* (4, 10). Further, species or serotype differentiation can also be achieved via multiplex PCR. Denis et al. (14) selected 16S *rRNA*, *mapA*, and *ceuE* as the target genes for simultaneous detection of *Campylobacter jejuni* and *Campylobacter coli*. A similar assay was established for simultaneous identification of *Salmonella* sp., *S.* Enteritidis, and *S.* Typhimurium in one reaction (64). In multiplex PCR, bacterial pathogens belonging to different genera can also be screened in the same amplification system. Li and Mustapha (41) and Li et al. (42) established a multiplex PCR for simultaneous detection of *E. coli* O157:H7, *Salmonella*, and *Shigella* in apple cider, produce and raw and ready-to-eat meat products. In most situations of multiplex PCR, the optimal conditions for different primer sets may be unique and interference among different primer pairs may occur, resulting in uneven amplification of different target sequences and limited sensitivity (68). Thus, adjusting the concentration of Taq DNA polymerase, $MgCl^2$, or dNTPs, as well as the concentration ratio among the different primer pairs is needed in the optimization of the amplification system. Although to design a robust multiplex PCR assay for foods can be challenging, once optimized for the specific pathogens and food products, this method has the advantage of being cost effective and highly efficient. Because of its selectivity, sensitivity, and efficiency, a multiplex PCR protocol is very applicable and suitable for comprehensive testings of specific foods.

REVERSE TRANSCRIPTION-PCR DETECTION OF FOODBORNE PATHOGENS

The use of reverse transcription PCR (RT-PCR) in foods is limited due to the difficulty of extracting undegraded mRNA from pathogens in complex food matrices. By amplifying the *iap* mRNA, a RT-PCR was successfully developed for detecting viable *L. monocytogenes* cells in cooked ground beef, artificially contaminated with ca. 3 cfu/g, following a 2-h enrichment step (34). McIngvale et al. (52) established a similar protocol for Shiga-toxin-producing *E. coli* with optimal growth medium, incubation temperature, and aeration. The assay was validated in artificially contaminated ground beef. Viable *E. coli* O157:H7 at an initial inoculum of 1 cfu/g was detectable in the meat after a 12-h enrichment. In addition, a RT-PCR was developed for detecting mRNA from the *sefA* gene of *S.* Enteritidis (66). The sensitivity of the assay depended on the physiological state of the cells under different temperatures and pH. With the RT-PCR, it was possible to detect 10 cells of *S.* Enteritidis PT4 in contaminated minced beef and whole egg samples following a 16-h enrichment step. Although not cost-effective for routine testings of pathogens in foods, these reports highlight the potential of RT-PCR for the detection of viable bacterial pathogens in foods.

REAL-TIME PCR DETECTION OF FOODBORNE PATHOGENS

Recent advances in fluorescent chemistries and detection instruments allow further development of PCR technology as a more efficient and sensitive tool for "real-time" microbiological analysis of foods. The use of nonspecific fluorescent double-stranded DNA-binding dyes (such as SYBRGreen or SYBRGold), or specific fluorescence resonance energy transfer technology (such as 5'-nuclease assay [TaqMan], or molecular beacon) has resulted in PCR assays with quantitative capability in a real-time manner (53). A number of real-time PCR assays have been described for the detection and quantification of *C. jejuni* (59), enterohemorrhagic *E. coli* serotypes O157, O111, and O26 in ground beef (62), *L. monocytogenes* in cabbage (25), *Salmonella* serotype D in egg (61), pathogenic *Vibrio* species in oyster (7, 50, 57), and *Y. enterocolitica* in ground pork (29, 72). Further, a sensitive multiplex real-time PCR has been developed for the simultaneous detection of *E. coli* O157:H7, *Salmonella,* and *Shigella* in pure culture and in ground beef (A. Mustapha, unpublished data). In addition to maintaining all the advantages of conventional PCR, real-time PCR has added speed and sensitivity. This technique can quantify a target DNA with greater reproducibility, which is very valuable in the quantitative assessment of microbial risks and the execution of HACCP programs in the food industry. The current drawback for using real-time PCR for routine food testing is the cost involved, not only in the equipment but the reagents.

CONCLUSIONS

The PCR has come a long way since its discovery, evolving from a tool used mainly in forensic, medical, pharmaceutical, and plant sciences to food science and the food industry. It may be one of the most remarkable discoveries of the 20th century and has opened new doors in a wide array of fields that would never have been possible prior to its utilization. Although research have shown that PCR can be a powerful method for detection of foodborne pathogens in pure culture as well as in certain foods, much more work needs to be done to truly make it the best alternative detection technique to conventional cultural methods. Until the enrichment steps can be eliminated, the rapidity of PCR assays can still be argued. Foods also are so different in their composition, resulting in a multitude of compounds that may be inhibitory to the detection of some pathogens while not affecting others, thus making it more challenging to design a one-size-fits-all PCR assay for foods.

REFERENCES

1. Aabo, S, O.F. Rasmussen, L. Rossen, P.D. Sorensen, and J.E. Olson. 1993. *Salmonella* identification by the polymerase chain reaction. Mol. Cell. Probes **7**:171–178.
2. Aabo, S., J.K Andersen, and J.E. Olson. 1995. Detection of *Salmonella* in minced meat by the polymerase chain reaction. Lett. Applied Microbiol. **21**:180–182.
3. Amagliani, G., G. Brandi, E. Omiccioli, A. Casiere, I.J. Bruce, and M. Magnani. 2004. Direct detection of *Listeria monocytogenes* from milk by magnetic based DNA isolation and PCR. Food Microbiol. **21**:597–603.
4. Becker, K., R. Roth, and G. Peters. 1998. Rapid and specific detection of toxigenic *Staphylococcus aureus*: Use of two multiplex PCR enzyme immunoassays for amplification and hybridization of staphylococcal enterotoxin genes, exfoliative toxin genes, and toxic shock syndrome toxin 1 gene. J. Clin. Microbiol. **36**:2548–2553.
5. Bolla, J.M., E. De, A. Dorez, and J.M. Pages. 2000. Purification, characterization, and sequence analysis of Omp50, a new porin isolated from *Campylobacter jejuni*. Biochem. J. **352**:637–643.
6. Bubert, A., S. Kohler, and W. Goebel. 1992. The homologous and heterologous regions within the *iap* gene allow genus- and species-specific identification of *Listeria* spp. by polymerase chain reaction. Appl. Environ. Microbiol. **58**:2625–2632.
7. Campbell, M.S. and A.C. Wright. 2003. Real-time PCR analysis of *Vibrio vulnificus* from oysters. Appl. Environ. Microbiol. **69**:7137–7144.
8. Cebula, T.A., W.L. Payne, and P. Feng. 1995. Simultaneous identification of strains of *Escherichia coli* serotype O157:H7 and their shiga-like toxin type by mismatch amplification mutation assay-multiplex PCR. J. Clin. Microbiol. **33**:248–250.
9. Chang, Y.H., Y.H. Shangkuan, H.C. Lin, and H.W. Liu. 2003. PCR assay of the *groEL* gene for detection and differentiation of *Bacillus cereus* group cells. Appl. Environ. Microbiol. **69**:4502–4510.
10. Chen, T.R., C.S. Chiou, and H.Y Tsen. 2004. Use of novel PCR primers specific to the genes of staphylococcal enterotoxin G, H, I for the survey of *Staphylococcus aureus* strains isolated form food-poisoning cases and food samples in Taiwan. Int. J. Food Microbiol. **92**:189–197.
11. Cocolin, L., M. Manzano, C. Cantoni, and G. Comi. 1998. Use of polymerase chain reaction and restriction enzyme analysis to directly detect and identify *Salmonella typhimurium* in food. J. Appl. Microbiol. **85**:673–677.
12. Davis, C.R., L.C. Heller, K.K. Peak, D.L. Wingfield, C.L. Goldstein-Hart, D.W. Bodager, A.C. Cannons, P.T. Amuso, and J.C. Cattani. 2004. Real-time PCR detection of the thermostable direct hemolysin and thermolabile hemolysin genes in a *Vibrio parahaemolyticus* cultured from mussels and mussel homogenate associated with a foodborne outbreak. J. Food Prot. **67**:1005–1008.
13. Dedieu, L., J.M. Pages, and J.M. Bolla. 2004. Use of the *omp50* gene for identification of *Campylobacter* species by PCR. J. Clin. Microbiol. **42**:2301–2305.
14. Denis, M., C. Soumet, K. Rivoal, G. Ermel, D. Blivet, G. Salvat, and P. Colin. 1999. Development of an m-PCR assay for simultaneous identification of *Campylobacter jejuni* and *C. coli*. Lett. Appl. Microbiol. **29**:406–410.
15. Ehling-Schulz, M., M. Fricker, and S. Scherer. 2004. Identification of emetic toxin producing *Bacillus cereus* strains by a novel molecular assay. FEMS Microbiol. Lett. **232**:189–195.
16. Ercolini, D., G. Blaiotta, V. Fusco, and S. Coppola. 2004. PCR-based detection of enterotoxigenic *Staphylococcus aureus* in the early stages of raw milk cheese making. J. Appl. Microbiol. **96**:1090–1096.

17. Fach, P. and M.R. Popoff. 1997. Detection of enterotoxigenic *Clostridium perfringens* in food and fecal samples with a duplex PCR and the slide latex agglutination test. Appl. Environ. Microbiol. **63**:4232–4236.

18. Fatemeh, R., H. Michael, H. Water, and C. Carl. 1995. Survival of *Shigella flexneri* on vegetables and detection by polymerase chain reaction. J. Food Prot. **58**:727–732.

19. Fluit, A.C., R. Torensma, M.J.C. Visser, C.J.M. Aarsman, J.J.G. Poppelier, B.H.I. Keller, P. Klapwijk, and J. Verhoef. 1993. Detection of *Listeria monocytogenes* in cheese with the magnetic immuno-polymerase chain reaction assay. Appl. Environ. Microbiol. **59**:1289–1293.

20. Fluit, A.C., M.N. Widjojoatmodjd, A.T.A. Box, R. Torensma, and J. Verhoef. 1993. Rapid detection of *Salmonella* in poultry with the magnetic immuno-polymerase chain reaction assay. Appl. Environ. Microbiol. **59**:1342–1346.

21. Fratamico, P.M., L.K. Bagi, and T. Pepe. 2000. A multiplex polymerase chain reaction of *Escherichia coli* O157:H7 in foods and bovine feces. J. Food Prot. **63**:1032–1037.

22. Fratamico, P.M. and T.P. Strobaugh. 1998. Simultaneous detection of *Salmonella* spp. and *Escherichia coli* O157:H7 by multiplex PCR. J. Industrial Microbiol. Biotechnol. **21**:92–98.

23. Harvard, H.L., E.C. Hunter, and W. Titball. 1992. Comparison of the nucleotide sequence and development of a PCR test for the epsilon toxin gene of *Clostridium perfringens* type B and type D. FEMS Microbiol. Lett. **97**:77–82.

24. Hansen, B.M., T.D. Leser, and N.B. Hendriksen. 2001. Polymerase chain reaction assay for the detection of *Bacillus cereus* group cells. FEMS Microbiol. Lett. **202**:209–213.

25. Hough, A.J., S.A. Harbison, M.G. Savill, L.D. Melton, and G. Fletcher. 2002. Rapid enumeration of *Listeria monocytogenes* in artificially contaminated cabbage using real-time polymerase chain reaction. J. Food Prot. **65**:1329–1332.

26. Houng, H.S., O. Sethabutr, and P. Echeverria. 1997. A simple polymerase chain reaction technique to detect and differentiate *Shigella* and enteroinvasive *Escherichia coli* in human feces. Diag. Microbiol. Infect. Dis. **28**:19–25.

27. Jackson, C.J., A.J. Fox, and D.M. Jones. 1996. A novel polymerase chain reaction assay for the detection and specification of thermophilic *Campylobacter* spp. J. Appl. Bacteriol. **81**:467–473.

28. Jitrapakdee, S., A. Tassanakajon, V. Boonsaeng, S. Piankijagum, and S. Panyim. 1995. A simple, rapid and sensitive detection of *Salmonella* in food by polymerase chain reaction. Mol. Cell. Probes **9**:375–382.

29. Jourdan, A.D., S.C. Johnson, and I.V. Wesley. 2000. Development of a fluorogenic 5′ nuclease PCR assay for detection of the *ail* gene of pathogenic *Yersinia enterocolitica*. Appl. Environ. Microbiol. **66**:3750–3755.

30. Jung, Y.S., J.F. Frank, R.E. Brackett, and J.R. Chen. 2003. Polymerase chain reaction detection of *Listeria monocytogenes* on frankfurters using oligonucleotide primers targeting the genes encoding internalin AB. J. Food Prot. **66**:237–241.

31. Karunasagar, I., G. Sugumar, I. Karunasagar, and P.J. Reilly. 1996. Rapid polymerase chain reaction method for detection of Kanagawa positive *Vibrio parahaemolyticus* in seafoods. Int. J. Food Microbiol. **31**:317–323.

32. Kim, S., R. G. Labbe, and S. Ryu. 2000. Inhibitory effects of collagen on the PCR for detection of *Clostridium perfringens*. Appl. Environ. Microbiol. **66**:1213–1215.

33. Kim, Y.B., J. Okuda, C. Matsumoto, N. Takahashi, S. Hashimoto, and M. Nishibuchi. 1999. Identification of *Vibrio parahaemolyticus* strains at the species level by PCR targeted to the *toxR* gene. J. Clin. Microbiol. **37**:1173–1177.

34. Klein, P.A. and V.K. Juneja. 1997. Sensitive detection of viable *Listeria monocytogenes* by reverse transcription-PCR. Appl. Environ. Microbiol. **63**:4441–4448.

35. Kwang, J., E.T. Littledike, and J.E. Keen. 1996. Use of the polymerase chain reaction for *Salmonella* detection. Lett. Appl. Microbiol. **22**:46–51.

36. Lampel, K.A., P.A. Orlandi, and L. Kornegay. 2000. Improved template preparation method for PCR-based assays for detection of foodborne bacterial pathogens. Appl. Environ. Microbiol. **66**:4539–4542.

37. Lantz, P.G., R.K. Knutsson, Y. Blixt, W.A. Al-Soud, E. Borch, and P. Radstrom. 1998. Detection of pathogenic *Yersinia enterocolitica* in enrichment media and pork by a multiplex PCR: A study of sample preparation and PCR-inhibitory components. Int. J. Food Microbiol. **45**:93–105.

38. Lee, J.W., I.J. Jun, H.J. Kwun, K.L. Jang, and J. Cha. 2004. Direct identification of *Vibrio vulnificus* by PCR targeting elastase gene. J. Microbiol. Biotechnol. **14**:284–289.

39. Lee, J.Y., Y.B .Bang, J.H. Rhee, and S.H. Choi. 1999. Two stage nested PCR effectiveness for direct detection of *Vibrio vulnificus* in natural samples. J. Food Sci. **64**:158–162.

40. Lett, P.W., J.P. Southworth, D.D. Jones, and A.K. Bej. 1995. Detection of pathogenic *Escherichia coli* in ground beef using multiplex PCR. Food Testing & Analysis **1**:34–38.

41. Li, Y. and A. Mustapha. 2004. Simultaneous detection of *Escherichia coli* O157:H7, *Salmonella*, and *Shigella* in apple cider and produce by a multiplex PCR. J. Food Prot. **67**:27–33.

42. Li, Y., S. Zhuang, and A. Mustapha. 2005. Application of a multiplex PCR for the simultaneous detection of *Escherichia coli* O157:H7, *Salmonella* and *Shigella* in raw and ready-to-eat meat products. Meat Sci. **71**:402–406.

43. Lin, C.K., C.L. Hung, S.C. Hsu, C.C. Tsai, and H.Y. Tsen. 2004. An improved PCR primer pair based on 16S rDNA for the specific detection of *Salmonella* serovars in food samples. J. Food Prot. **67**:1335–1343.

44. Lin, J.S. and H.Y. Tsen. 1999. Development and use of polymerase chain reaction for the specific detection of *Salmonella* Typhimurium in stool and food samples. J. Food Prot. **62**:1103–1110.

45. Lindqvist, R. 1999. Detection of *Shigella* spp. in food with a nested PCR method—Sensitivity and performance compared with a conventional culture method. J. Appl. Microbiol. **86**:971–978.

46. Linton, D., A.J. Lawson, R.J. Owen, and J. Stanley. 1997. PCR detection, identification to species level, and fingerprinting of *Campylobacter jejuni* and *Campylobacter coli* direct from diarrheic samples. J. Clin. Microbiol. **35**:2568–2572.

47. Liu, D., A.J. Ainsworth, F.W. Austin, and M.L. Lawrence. 2004. Use of PCR primers derived from a putative transcriptional regulator gene for species-specific determination of *Listeria monocytogenes*. Int. J. Food Microbiol. **91**:297–304.

48. Longhi, C., A. Maffeo, M. Penta, G. Petrone, L. Seganti, and M.P. Conte. 2003. Detection of *Listeria monocytogenes* in Italian-style soft cheeses. J. Appl. Microbiol. **94**:879–885.

49. Luk, J.M., U. Kongmuang, P.R. Reeves, and A.A. Lindberg. 1993. Selective amplification of abequose and paratose synthase gene (*rfb*) by polymerase chain reaction of *Salmonella* major serotypes (A, B, C2, D). J. Clin. Microbiol. **31**:2118–2123.

50. Lyon, W.J. 2001. Taqman PCR for detection of *Vibrio cholerae* O1, O139, Non-O1, and Non-O139 in pure cultures, raw oysters, and synthetic seawater. Appl. Environ. Microbiol. **67**:4685–4693.

51. Manzano, M., C. Giusto, L. Lacumin, C. Cantoni, and G. Comi. 2003. A molecular method to detect *Bacillus cereus* from a coffee concentrate sample used in industrial preparations. J. Appl. Microbiol. **95**:1361–1366.

52. McIngvale, S.C., D. Elhanafi, and M.A. Drake. 2002. Optimization of reverse transcriptase PCR to detect viable Shiga-toxin-producing *Escherichia coli*. Appl. Environ. Microbiol. **68**:799–806.

53. McKillip, J.L. and M. Drake. 2004. Real time nucleic acid-based detection methods for pathogenic bacteria in food. J. Food Prot. **67**:823–832.

54. Meng, J., S. Zhao, M.P. Doyle, S.E. Mitchell, and S. Kresovich. 1997. A multiplex PCR for identifying Shiga-like toxin-producing *Escherichia coli* O157:H7. Lett. Appl. Microbiol. **24**:172–176.

55. Miyamoto, T., N. Ichioka, C. Sasaki, H.K. Kobayashi, K. Honjoh, M. Iio, and S. Hatano. 2002. Polymerase chain reaction assay for detection of *Escherichia coli* O157:H7 and *E. coli* O157:H⁻. J. Food Prot. **65**:5–11.

56. Osek, J. 2001. Multiplex polymerase chain reaction assay for identification of enterotoxigenic *Escherichia coli* strains. J. Vet. Diag. Invest. **13**:308–311.

57. Panicker, G., M.L. Myers, and A.K. Bej. 2004. Rapid detection of *Vibrio vulnificus* in shellfish and Gulf of Mexico water by real-time PCR. Appl. Environ. Microbiol. **70**:498–507.

58. Rasmussen, H.N., J.E. Olsen, K. Jorgensen, and O.F. Rasmussen. 1996. Detection of *Campylobacter jejuni* and *Campylobacter coli* in chicken faecal samples by PCR. Lett. Appl. Microbiol. **23**:363–366.

59. Sails, A.D., A.J. Fox, F.J. Bolton, D.R. Waring, and D.L. Greenway. 2003. A real-time PCR assay for the detection of *Campylobacter jejuni* in foods after enrichment culture. Appl. Environ. Microbiol. **69**:1383–1390.

60. Schraft, H. and M.W. Griffiths. 1995. Specific oligonucleotide primers for detection of lecithinase positive *Bacillus* spp. by PCR. Appl. Environ. Microbiol. **61**:98–102.

61. Seo, K.H., I.E. Valentin-Bon, R.E. Brackett, and P.S. Holt. 2004. Rapid, specific detection of *Salmonella* Enteritidis in pooled eggs by real-time PCR. J. Food Prot. **67**:864–869.

62. Sharma, V.K. 2002. Detection and quantitation of enterohemorrhagic *Escherichia coli* O157, O111, and O26 in beef and bovine feces by real-time polymerase chain reaction. J. Food Prot. **65**:1371–1380.

63. Somer, L. and Y. Kashi. 2003. A PCR method based on 16S rRNA sequence for simultaneous detection of the genus *Listeria* and the species *Listeria monocytogenes* in food products. J. Food Prot. **66**:1658–1665.

64. Soumet, C., G. Ermel, N. Rose, V. Drouin, G. Salvat, and P. Colin. 1999. Evaluation of a multiplex PCR assay for simultaneous identification of *Salmonella* sp., *Salmonella* Enteritidis, and *Salmonella* Typhimurium from environmental swabs of poultry houses. Lett. Appl. Microbiol. **28**:113–117.

65. Straub, J.A., C. Hertel, and W.P. Hammes. 1999. A 23S rDNA-targeted polymerase chain reaction-based system for detection of *Staphylococcus aureus* meat starter cultures and dairy products. J. Food Prot. **62**:1150–1156.

66. Szabo, E.A., and B.M. Mackey. 1999. Detection of *Salmonella* Enteritidis by reverse transcription-polymerase chain reaction. Int. J. Food Microbiol. **51**:113–122.

67. Tamarapu, S., J.L. McKillip, and M. Drake. 2001. Development of a multiplex polymerase chain reaction assay for detection and differentiation of *Staphylococcus aureus* in dairy products. J. Food Prot. **64**:664–668.

68. Taylor, M.S., A. Challed-Spong, and E.A. Johnson. 1997. Co-amplification of the amelogenin and HLA DQ? genes: Optimization and validation. J. Forensic Sci. 42:130–136.
69. Tsen, H.Y., L.Z. Jian, and W.R. Chi. 1998. Use of a multiplex PCR system for the simultaneous detection of heat labile toxin I and heat stable toxin II genes of enterotoxigenic *Escherichia coli* in skim milk and porcine stool. J. Food Prot. 61:141–145.
70. Uzal, F.A., J.J. Plumb, L.L. Blackall, and W.R. Kelly. 1997. PCR detection of *Clostridium perfringens* producing different toxins in faeces of goats. Lett. Appl. Microbiol. 25:339–344.
71. Villalobo, E. and A. Torres. 1998. PCR for detection of *Shigella* spp. in mayonnaise. Appl. Environ. Microbiol. 64:1242–1245.
72. Vishnubhatla, A., R.D. Oberst, D.Y. Fung, W. Wonglumsom, M.P. Hays, and T.G. Nagaraja. 2001. Evaluation of a 5′ nuclease (TaqMan) assay for the detection of virulent strains of *Yersinia enterocolitica* in raw meat and tofu samples. J. Food Prot. 64:355–360.
73. Wilson, I.G. 1997. Inhibition and facilitation of nucleic acid amplification. Appl. Environ. Microbiol. 63:3741–3751.
74. Yeung, P.S., A. Depaola, C.A. Kaysner, and K.J. Boor. 2003. A PCR assay for specific detection of the pandemic *Vibrio parahaemolyticus* O3:K6 clone from shellfish. J. Food Sci. 68:1459–1466.

CHAPTER 6

Molecular Approaches for the Detection of Foodborne Viral Pathogens

Doris H. D'Souza[*] and Lee-Ann Jaykus

*Department of Food Science, College of Agriculture and Life Sciences,
North Carolina State University, Raleigh, NC 27695-7624*

Introduction
General Detection Considerations and the Challenges
Virus Concentration
  Virus Concentration Methods for Shellfish
  Virus Concentration Methods for Other Than Shellfish
Nucleic Acid Extraction
Detection
  RT-PCR Detection of Viruses in Foods
  Alternative Nucleic Acid Amplification Methods
Confirmation
Real-Time Detection
Conclusions
Acknowledgments
References

INTRODUCTION

The human enteric viruses are now recognized as major causes of acute non-bacterial gastroenteritis throughout the world. Those of primary epidemiological significance include hepatitis A virus (HAV) and the noroviruses, formerly known as the Norwalk-like viruses (NLVs), or the small round structured viruses (SRSVs) (reviewed in 88). Currently, the noroviruses consist of five genogroups: GI (prototype Norwalk virus); GII (prototype Snow Mountain Agent); GIII (prototype bovine enteric calicivirus); GIV (prototype Alphatron and Fort Lauderdale virus); and GV (prototype Murine norovirus) (Vinje, personal communication). The sapoviruses (previously called Sapporo-like viruses) are genetically related to the noroviruses and have occasionally caused viral gastroenteritis in humans. Both the noroviruses and the sapoviruses are members of the *Calicivirideae* family, an antigenically and genetically diverse group of gastrointestinal viruses (9, 39, 40). Other viruses that can cause food and waterborne disease include the adenoviruses, astroviruses, the human enteroviruses (polioviruses, echoviruses, groups A and B coxsackieviruses), hepatitis E virus,

[*] Corresponding author: Phone: (919) 513-2076, FAX: (919) 513-0014 e-mail: ddsouza@unity.ncsu.edu

parvoviruses, and other relatively uncharacterized small round viruses. The rotaviruses, which are the leading cause of infantile diarrhea worldwide, are transmitted primarily by contaminated water but can on occasion be foodborne.

The human enteric viruses replicate in the intestines of infected human hosts and are excreted in the feces. Their primary mode of transmission is the fecal–oral route through contact with human fecal matter, although they may also be shed in vomitus. These viruses are readily spread by person-to-person contact, which is frequently responsible for the propagation of primary food-borne outbreaks. Contamination of foods may occur directly, through poor personal hygiene practices of infected food handlers, or indirectly via contact with fecally contaminated waters or soils (49, 50). Since viruses must survive the pH extremes and enzymes present in the human gastrointestinal tract, they are regarded as highly environmentally stable, allowing virtually any food to serve as a vehicle for their transmission (50). Although enteric viruses are unable to replicate in contaminated foods, they are able to withstand a wide variety of food processing and storage conditions. When present in contaminated food, their numbers are usually quite low, but since their infectious doses are also low, any level of contamination may pose a public health threat.

In spite of their initial recognition decades ago, the human enteric viruses can be considered "emerging" agents of foodborne disease, mainly because only recently have scientists been able to reliably detect these pathogens. In fact, prior to the advent of molecular biological techniques, epidemiological criteria were the best means, by which cases of enteric viral illness were recognized. Over the last 10 years, significant advances in nucleic acid amplification methods have made detection of enteric viruses in human clinical samples, specifically feces, all but routine.

The opposite is the case for the detection of viruses in foods. Historically, this has been done by infectivity assay using susceptible, live laboratory hosts. Host systems employed were mainly mammalian cell cultures of primate origin. Unfortunately, the epidemiologically important human enteric viruses, including the noroviruses and wild-type hepatitis A virus, cannot be propagated in mammalian cell culture systems, and so these are not viable detection options (49, 50). In the absence of *in vitro* virus propagation methods, nucleic acid amplification has been a promising alternative. In this chapter, we will discuss existing technologies that can be applied to the detection of viruses in foods, and we will address new developments and research needs for the application of these methods on a routine basis.

GENERAL DETECTION CONSIDERATIONS AND THE CHALLENGES

The development of effective virus detection methods from food commodities poses several challenges. Like many bacterial pathogens, these agents are typically present at low levels in contaminated foods. However, unlike bacterial pathogens, viruses cannot replicate in foods, making the use of traditional food microbiological techniques of cultural enrichment and selective plating inapplicable *per se*. Therefore, the first goal in developing virus detection methods for foods is to separate and concentrate the agents from the food matrix. It is also necessary to sample relatively large volumes to assure adequate sample representation, thereby optimizing detection assay sensitivity.

VIRUS CONCENTRATION

Sample preparation prior to detection is of key importance when applying molecular methods to detect viral contamination in foods. In this regard, specific challenges include high sample volumes in relation to small amplification volumes, low levels of contamination, and the presence of residual food components that can later compromise detection (31, 37, 86, 97).

The main goals of virus concentration methods are to decrease sample volume and eliminate matrix-associated interfering substances, while simultaneously recovering most of the viruses present in the food sample. In order to achieve these goals, sample manipulations are undertaken that capitalize on the behavior of viruses to act as proteins in solutions, and to remain infectious at extremes of pH or in the presence of organic solvents. Because of their frequent association with viral foodborne disease outbreaks, early work on virus concentration and purification from foods focused mainly on bivalve molluscan shellfish. Recent research endeavors have included a broader range of at-risk foods.

Two major approaches to virus concentration, particularly as applied to shellfish commodities, are termed extraction–concentration and adsorption–elution–concentration (49). Both methods utilize conditions that promote the separation of viruses from shellfish tissues, through the use of filtration, centrifugation, adsorption, elution, solvent extraction, precipitation, and/or organic flocculation. The procedure generally begins with sample blending in a buffer, usually containing amino acids and an elevated pH. A common example of elution buffer is 0.1 M glycine-0.14 N saline, pH 9.0. A crude filtration step through a mesh material such as cheesecloth may be done to remove large sample particulates. Viruses do not sediment unaided, even at centrifugation speeds approaching 10,000 × g. Therefore, centrifugation can be used to sediment large food particles, with the recovery of the virus-containing supernatant. The next step usually involves pH manipulation or the addition of precipitation agents, creating conditions such that viruses adsorb to the remaining shellfish tissues. Upon subsequent centrifugation, the adsorbed viruses sediment with the tissues and the supernatant is discarded. Elution, whereby the viruses are desorbed from the tissues by further pH and/or ionic manipulations is then carried out. On subsequent centrifugation, the precipitated tissue is discarded, again retaining the supernatant. Using sequential steps of adsorption, elution, filtration, precipitation, and centrifugation, viruses are concentrated to small sample volumes and simultaneously purified, with the removal of large proportions of the food matrix and matrix-associated organic materials that may later compromise detection of the viruses.

Precipitation of viruses can be achieved by lowering pH, so called acid precipitation, or by the addition of polyethylene glycol (PEG). Both methods capitalize on the property that viruses behave as proteins in solution. The viruses, along with some of the matrix-associated proteins, will precipitate when the pH is lowered to that approximating the virus isoelectric point. Polyethylene glycol causes removal of water, allowing proteins to fall out of solution. Another method similar to precipitation is organic flocculation. Flocculating agents interact with organic material in the matrix, causing the formation of a gelatinous "floc," to which the viruses absorb (49, 50). In the case of acid and PEG precipitation, and in organic flocculation, the virus-containing solid materials can be readily harvested by centrifugation, usually done at fairly low speeds (e.g., <5000 × g).

Further removal of matrix-associated organic materials can be done using a variety of agents. Since viruses remain infectious even after exposure these organic solvents such as chloroform, trichloro trifluoroethane (Freon) and more environmentally friendly solvents such as Vertrel (Dupont), these chemicals can be used to remove polar food components such as lipids. Alternative commercial virus purification agents such as ProCipitate and Viraffinity (LigoChem, Inc., Fairfield, NJ 07004) (31, 51, 66) are known to eliminate polysaccharides, an important matrix-associated inhibitor in shellfish and produce. The cationic detergent cetyltrimethylammonium bromide (CTAB) (6, 7, 52) also aids in the removal of polysaccharides. The use of Sephadex (23), cellulose (110), or Chelex (98) is helpful in the elimination of salts and small proteins. Ultrafiltration, a method that is frequently applied at the latter stages of a virus concentration scheme, provides reduction in volume while simultaneously purifying the sample.

VIRUS CONCENTRATION METHODS FOR SHELLFISH

The early molecular work aimed at detecting viruses in the food matrix focused almost exclusively on shellfish. The most common method utilizes some sort of virion concentration step prior to release or extraction of nucleic acid, followed by amplification. A second, less commonly used method resorts to direct nucleic acid extraction of a previously untreated food matrix, which circumvents the need for virus concentration.

The objective of the virion concentration approach is to concentrate viruses and remove inhibitors prior to nucleic acid amplification, with or without prior nucleic acid extraction. In early work by Atmar et al. (6, 7), investigators processed artificially contaminated shellfish samples using an initial concentration and utilized a purification scheme that consisted of solvent extraction and PEG precipitation steps. This was followed by nucleic acid extraction and subsequent amplification. Cetyltrimethylammonium bromide (CTAB) was added to remove residual inhibitors from crude nucleic acid extracts, and the resulting eluant was amplified using reverse transcriptase-polymerase chain reaction (RT-PCR) (6, 7). Dissecting the oysters, discarding the muscle tissue and processing only the digestive diverticula improved the PCR's detection limits. This sampling approach is now the method of choice (11, 65, 67, 69, 90, 92). Figure 6.1 illustrates a representative sample-preparation protocol for shellfish. Note that second generation protocols frequently employ sequential PEG precipitation in addition to adsorption, elution, and solvent extraction steps (8, 16, 19, 20, 25, 26, 30, 59, 60, 63, 64).

Some investigators apply an antibody capture step to further concentrate and purify viruses from shellfish extracts prior to detection using RT-PCR.

25 g OYSTER MEAT
↓
HOMOGENIZATION
Add 175 ml of sterile cold deionized water (ratio of meat: water is 1:7)
↓
VIRUS ADSORPTION
Adjust pH to 4.8 and with conductivity of < 2000 µS using water
↓
CENTRIFUGATION
Collect pellets at 2,000 x g for 20 min
↓
VIRUS ELUTION
Add 0.75 M glycine-0.15 M NaCl, pH 7.6 (1:7 ratio of meat: eluant buffer)
pH adjustment to 7.5 to 7.6, Vortex at room temp for 15 mins
Centrifuge at 5,000 x g for 20 min at 4 C
Collect supernatant (A)
↓
RE-ELUTE VIRUS FROM PELLET
Add 0.5 M threonine-0.15 M NaCl, pH 7.6 (1:7 ratio)
Collect supernatant (B)
Combine/Pool supernatant A and B
↓
PEG PRECIPITATION
Add 8% PEG 8000-0.3 M NaCl
Incubate at 4 C for 4 h or overnight
Centrifuge 6,700 x g for 30 min
Collect pellets and resuspend in10 ml of Phosphate-buffered saline
↓
SOLVENT EXTRACTION OF VIRUS
Add 10 ml of chloroform (ratio of eluant: chloroform is 1:1)
Centrifuge at 1, 700 x g for 30 min
Collect supernatant 1
↓
Reextract from bottom layer with half volume of 0.5 M threonine
Collect supernatant 2
Combine/pool supernatant 1 and 2
↓
PEG PRECIPITATION
Add 8% PEG-0.3 M NaCl at 4 C for 2-4 h
Centrifuge at 14,000 x g for 15 min, Collect pellets
↓
RNA EXTRACTION

Figure 6.1. Concentration of viruses from oysters using virus adsorption, glycine-saline buffer and threonine-saline extraction, PEG precipitation, chloroform extraction, and PEG concentration as modified by Shieh et al. (95).

Desenclos et al. (29) were the first to implicate hepatitis A virus in oyster out-break specimens by immunocapture of the virus, heat release of viral nucleic acids, and subsequent RT-PCR detection. Capitalizing on the work of Jansen et al. (48), other investigators have coated paramagnetic beads with anti-HAV IgG and used these to capture HAV from oyster extracts initially processed for virus concentration using a combination of elution, polyelectrolyte floccula-tion, filtration and/or ultrafiltration (28, 71). Sunen et al. (100) and Schwab et al. (94) also used antibody capture as a final virus concentration step. More recently, Kobayashi et al. (61) used magnetic beads coated with the antibody to the baculovirus-expressed recombinant capsid proteins of the Chiba virus (rCV) to capture noroviruses from food items implicated in an outbreak of acute gastroenteritis in Aichi Prefecture, Japan, detecting the virus in these foods by RT-PCR. Abd El Galil et al. (1) have used immunomagnetic separa-tion for the detection of hepatitis A virus from environmental samples using real-time nucleic acid amplification methods.

The alternative approach of direct nucleic acid extraction and RT-PCR applied to unprocessed food sample, involves extraction of total RNA from the sample without any prior sample manipulations. This method is best suited for simple sample matrices such as the surfaces of fresh fruits and vegetables. However, Legeay et al. (65) recently reported a method that involved enzymatic liquefaction of shellfish digestive tissues, followed by clarification using dichloromethane extraction. In this case, the investigator reported that the sam-ple could be directly processed for nucleic acid isolation and subsequent virus detection by RT-PCR.

VIRUS CONCENTRATION METHODS FOR FOODS OTHER THAN SHELLFISH

Gouvea et al. (36) first reported a systematic method for the detection of norovirus and rotavirus from representative food commodities other than shell-fish, including orange juice, milk, lettuce, and melon. The method involved blending or washing, clarification by centrifugation, and removal of inhibitors by Freon extraction followed by RNA extraction. Leggitt and Jaykus (66) devel-oped a prototype method for the concentration of poliovirus, hepatitis A virus, and Norwalk virus from 50 g samples of artificially contaminated hamburger and lettuce. The steps used included homogenization, filtration through cheese-cloth, Freon extraction (hamburger only) and two sequential PEG precipita-tions. The sequential precipitations, which used increasing PEG concentrations, resulted in a 10- to 20-fold sample volume reduction from 50 g to approximately 2.5 ml (66). The resuspended PEG precipitate could be assayed for virus recov-ery by mammalian cell culture infectivity assay, when applicable, which allows for direct comparison between virus infectivity and molecular detection (104). Subsequent nucleic acid extraction resulted in an additional 100-fold sample volume reduction with detection at initial inoculum levels of $\geq 10^2$ infectious units per 50-g food sample (66). A schematic overview of this procedure is provided in Figure 6.2. Schwab et al. (91) used TRIzol, a proprietary RNA extraction method, as a surface wash for deli meats, including samples artificially contaminated with norovirus and ones implicated in an outbreak of norovirus gastroenteritis. Although simple, the main drawback of the TRIzol surface wash method was that nucleic acid amplification inhibition persisted unless sample concentrates were diluted 10- to 100-fold. A flow diagram of this protocol is depicted in Figure 6.3.

Bidawid et al. (12) reported an immunocapture method for the concentration of hepatitis A virus from lettuce and strawberries. After surface washing to elute the viruses, the wash solution was passed through a positively charged filter, eluted, and concentrated by immunocapture. As few as 10 PFU of cell culture-adapted hepatitis A virus per piece of lettuce or strawberry could be detected by RT-PCR using this sample preparation method. However, this method too had problems with residual matrix-associated amplification inhibition.

50 g FOOD SAMPLE
(Complex food such as hamburger)
Add 350 ml of 0.05 M Glycine/0.14 N Saline, pH 9.0
HOMOGENIZE (350-400 ml)

↓

FILTER
Cheesecloth

↓

SOLVENT EXTRACT
(Vortex gently with 60% Freon or Chloroform: Isobutanol (1:1) as needed, ~70 ml)
Centrifuge at 2,500 x g for 10 min.

↓

1° PEG PRECIPITATION and ELUTION
To eluant add 6% PEG, pH to 7.2, incubate 4°C, 2 H.
Centrifuge 5000 xg for 15 min.
Resuspend pellet in 25 ml of 50mM Tris, 0.2% Tween 20, pH 9.0
Elute at room temperature for 1 H.
Centrifuge at 3500 x g for 15 min.
(30-35 ml)

↓

2°PEG PRECIPITATION and RESUSPENSION
To eluant add 12% PEG; incubate 4°C, 2 H.
Centrifuge at 5000 x g for 15 min.
Resuspend pellet in 50mM Tris, 0.2% Tween 20, pH 8.0
(3-5 ml)

↓   ↓

RNA EXTRACTION and RT-PCR   OR   CELL-CULTURE ASSAY
(25-40 µl)

Figure 6.2. Virus concentration from hamburger/complex foods using glycine-saline buffer, chloroform extraction and two PEG extractions as followed by Leggitt and Jaykus (66)

25-50 g PRODUCE SAMPLE

TRIZOL® SURFACE WASH   OR   BUFFERED SURFACE WASH
(4 mls of TRIzol® Reagent, twice)   0.05 M Glycine-0.14M Saline buffer, pH 9.0
(4 ml of buffer, twice)

ELUTE AND CLARIFY SUPERNATANT
Centrifuge 8ml at 5000 rpm for 15 mins at 4°C

EXTRACT RNA
Add 1.6 ml of Chloroform to 8 ml of TRIzol/Glycine extract
Shake and incubate at room temperature for 2-3 min.

Centrifuge 9000 rpm for 15 min. at 4°C
Collect aqueous phase (~ 4.8 ml)

Add 4 ml of isopropanol and incubate at room temperature for 10 min.

Centrifuge at 9000 rpm for 10 min. at 4°C
Wash pellet with 8ml of 75% ethanol
Centrifuge 7000 rpm for 5 min. at 4°C

Air-dry for 5 min.
Dissolve pellet in 100 µl of sterile water and store at –70°C to –80°C

Figure 6.3. Virus concentration and extraction from produce using TRIzol® or glycine-saline buffer (modifications of protocol by Schwab et al. (91)).

As a general rule, virus concentration methods result in sample volume reductions ranging from 10- to 1000-fold. This means that a 25-g sample theoretically can be reduced to 25 µl–2.5 µl volumes with recovery of infectious virus (50). The yields after virus concentration in a food matrix have ranged from as low as 1–2% to as high as 90%. Recovery efficiency is almost always virus-specific, usually hepatitis A virus recovery is quite low when compared to recovery of other viruses such as poliovirus (reviewed in 50, 88).

Although virus concentration and sample purification steps prior to nucleic acid amplification can achieve significant sample volume reduction with relatively efficient virus recovery, there are some additional considerations when selecting a virus concentration approach. The antibody capture methods are often simpler than others as they may require fewer sample manipulations. There is also speculation that antigen-associated viral nucleic acid is more highly associated with infectious virus. However, these methods may be limited by reagent availability and high specificity, which means that only a single virus type is detected in a single assay. Other concentration and purification methods that rely on steps such as PEG precipitation and solvent extraction usually require manipulations, which may result in substantial virus loss during extraction. The direct nucleic acid extraction methods almost always result in residual RT-PCR inhibitors and often times do not provide adequate sample volume reduction.

In short, there are many virus extraction and concentration approaches that have been applied in a variety of instances. All of these methods have limitations that ultimately hinder the ability to detect the relatively low levels of virus that might be anticipated in naturally contaminated foods.

NUCLEIC ACID EXTRACTION

The earliest and simplest protocol for amplifying nucleic acids from shellfish concentrates was direct heat release followed by RT-PCR, in which case, an RNA extraction step was not applied (31, 51, 52). It soon became apparent, however, that in many instances the additional volume reductions and sample clean up provided by RNA extraction was critical to improving detection limits and circumventing the effects of residual matrix-associated amplification inhibition.

Before applying nucleic acid amplification, an efficient nucleic acid extraction step is critical in most instances. This is important because the amplification efficiency is dependent on both the purity of the target template and the quantity of target molecules obtained from the sample. Accordingly, the main goals of nucleic acid extraction are: (1) to extract and purify the nucleic acids, (2) to provide additional sample concentration, and (3) to remove residual matrix-associated inhibitory substances that could remain after the initial concentration steps are completed. The components that can hinder molecular amplification are diverse and include compounds such as divalent cations, matrix-associated components such as proteoglycans (42), polysaccharides, (6, 7, 27), glycogen, (6, 7, 51, 52), and lipids (84, 87, 91), among others. In many cases, inhibitory compounds have been largely uncharacterized. Matrix-associated inhibitors usually act by degrading the target and/or primer nucleic acids, and/or inactivating or inhibiting enzymes (86, 109, 111). Unfortunately, these inhibitory substances are frequently coextracted during virus concentration protocols, adding to the challenge of identifying reliable nucleic acid extraction processes.

Early studies used SDS-proteinase K digestion to release nucleic acids from shellfish concentrates, followed by phenol chloroform extraction with, or without the addition of cetyltrimethylammonium bromide (CTAB) to remove residual inhibitors (6, 7, 16, 90). In the last decade or so, guanidinium thiocyanate (GuSCN)-based methods became the RNA extraction method of choice largely because they are effective at deproteinization of nucleic acids while providing ample protection of RNA against native RNases. Many commercial guanidinium based kits have been used in more recent studies (4, 12, 22, 24, 25, 32, 59, 60, 75, 87, 90, 91).

Combinations of multiple extraction methods can also be used to purify nucleic acids. A simple and rapid protocol for the purification of nucleic acid that utilizes a combination of the chaotropic agent GuSCN and silica particles was first described by Boom et al. (13), and later used by others (41, 53). Other methods include a GuSCN method followed by RNA binding to glass powder, instead of silica, to provide further nucleic acid purification (63, 64).

Several investigators have compared various RNA extraction approaches specifically aimed at preparing samples for the detection of human enteric viruses. Hale et al. (44) compared four different RNA extraction methods for RT-PCR detection of noroviruses in fecal specimens, finding that the GuSCN/silica (13) method was the best at removal of inhibitors. Likewise, another study compared seven RNA extraction methods to purify hepatitis A virus RNA from stool and shellfish concentrates for RT-PCR detection (4); again, the GuSCN-silica methods were found to be the most suitable from the standpoint of speed, ease, and cost. Gouvea et al. (36) used deproteinization with GuSCN followed by adsorption of RNA to hydroxyapatite and sequential precipitation with CTAB and ethanol to purify RNA from shellfish and other selected foods. Sair et al. (87) compared multiple RNA extraction methods after concentration of noroviruses from model food commodities (hamburger sandwiches and lettuce). These included GuSCN, commercial microspin columns, the QIAshredder Homogenizer and TRIzol alone and in their various combinations. These investigators found that the use of TRIzol followed by further sample preparation using the QIAshredder Homogenizer yielded the best detection limits (<1 RT-PCR amplifiable units/reaction) for Norwalk virus preconcentrated from food samples. Similar studies by Svensson (101) demonstrated that the use of the metal chelating agent Chelex-100, or alternatively, Sephadex G200 column chromatography, during RNA extraction, provided the best RT-PCR detection limits. Others have had success using phenol-chloroform-based methods followed by further selection for viral RNA using magnetic poly (dT) beads (30, 35, 59, 60).

Despite all the efforts in identification of efficacious RNA extraction protocols, food-related amplification inhibitors frequently remain. Multiple sample manipulation steps can result in incomplete recovery and/or degradation of RNA during the extraction procedure, the consequence of which is less than optimal RNA yields (49, 50). A major problem with RNA extraction is the necessity to destroy virion integrity, thereby losing the ability to directly correlate infectivity to RT-PCR detection limits, at least when effective cell culture-based methods are available. The two main areas of active research in RNA purification are increasing yield and improving the purity of the resultant product for detection by PCR.

DETECTION

The progress in clinical detection of pathogens has always been ahead of detection in foods and many of our food methods rely on protocols initially developed for clinical samples. However, whether clinical, food, or environmental sample, the sensitivity and the specificity of molecular amplification methods is largely dependent on the choice of primers.

The genetic diversity in the *Calicivirideae* family makes primer design for the detection of the noroviruses quite challenging. Initial studies used extremely specific primers such as NV 5′/3′ and NV 36/35, which were based on sequences in the prototype Norwalk virus genome (23, 74). With the availability of sequence information from related viruses, more broadly reactive primers have been designed. (2, 38, 57, 108). Most of the primer sequences reported are based on the highly conserved RNA-dependent RNA polymerase region of the noroviruses; occasionally the capsid region has been targeted (39, 40, 62, 81, 105).

Initially, the primer sets developed by Ando et al. (2) for the genogroup I (GI) and genogroup II (GII) Noroviruses were used as the "gold-standard" for the detection of noroviruses in clinical (fecal) and food samples (47, 80, 112). Later on, degenerate primers, or a mixture of oligonucleotide primers that vary in nucleotide sequence but have the same number of nucleotides, were developed and used for the detection of the noroviruses (38, 68, 69). The use of degenerate primers is advantageous in that all combinations of nucleic acid sequence that code for the amino acid are used in the PCR amplification. Combinations of the Ando et al. (2) GI and GII primers and various degenerate sets are routinely used in norovirus epidemiological investigations as applied to the detection of virus in fecal samples.

As with RNA extraction methods, investigators have compared the performance of various primers for the detection of a broad range of noroviruses, largely in fecal specimens. For instance, the NV110/NV36 primer set was the found to be the most efficient of the nine primer sets tested in a comprehensive study, even though it could not detect 100% of the norovirus strains tested (46). When five laboratories in five countries evaluated different RT-PCR methods on a panel of 91 fecal samples (106), no single assay was superior based on the criteria of sensitivity, detection limit, assay format, and successful implementation. However, the Boom extraction method and the use of the JV12/JV13 primer set were recommended for norovirus diagnostics.

Nonetheless, there is some lack of consensus regarding the optimal primer pair(s) to detect the noroviruses and different laboratories tend to use various methods that were developed, or are optimally suited for their purposes (112). Primers used for the detection of hepatitis A virus usually target the VP1/2A junction sequence. In this case, the issue of diversity is not relevant as it is for the noroviruses (48).

RT-PCR DETECTION OF VIRUSES IN FOODS

The choice of primers is even more critical when attempting to detect viral contamination in foods. This is because the levels of contamination are typically much lower in foods when compared to clinical samples, and even with optimal concentration and nucleic acid extraction methods, residual inhibitors often persist. Furthermore, the matrix can be responsible for nonspecific amplification and false positive results. The primers selected for the detection of viral nucleic acids derived from the food matrix should therefore have the following criteria: (1) a reasonably high annealing temperature, (2) relative nondegeneracy, and (3) broad reactivity. High stringency and primer specificity (hence the relative absence of degeneracy) are necessary to prevent nonspecific amplification. For the genetically diverse norovirus group, the use of primers that are broadly reactive and can detect as many genetically distinct strains as possible in a single assay is essential.

The various primers used in the RT-PCR detection of noroviruses from different food matrices are summarized in Table 6.1. As with primers used in the clinical realm, these sequences correspond to mainly the viral RNA dependent RNA polymerase or the capsid protein genome regions. Primer sets targeting both the RNA polymerase and capsid genes, Mon381/383 and SR33/46, respectively, were used by Shieh et al. (96, 97) to identify a GII norovirus in oyster samples implicated in a California outbreak. For the detection of the GII noroviruses in shellfish, the NI/E3 primer set has also been used (33, 37, 63). Dubois et al. (32) used a newer primer pair to detect both noroviruses and sapoviruses in artificially contaminated produce. In a systematic comparison of four primer pairs, as applied to the detection of noroviruses in hamburger sandwiches and lettuce, Sair et al. (87) found the best detection limits using the NVp110/NVp36 primer combination. Honma et al. (46) reported that this same primer pair was broadly reactive for a range of noroviruses, eliminating the need for separate amplifications for the two norovirus genogroups (GI and GII). These primers have also been used together or in combination with NI or NVp69 for the detection of norovirus contamination in shellfish (59, 67, 69).

"Nested" RT-PCR or double amplification has been used for the detection of noroviruses (36, 63, 91, 99) and hepatitis A virus (18, 19, 20, 43, 59) in foods. This approach can improve assay sensitivity and also provide another method for the confirmation of amplified product. However, a major disadvantage is that these nested reactions are prone to carryover contamination. Novel single-tube nested RT-PCR methods may help circumvent these issues. Ratcliff et al. (81) pooled the reagents required for the nested amplification in a "hanging drop" that could be introduced by centrifugation after the first RT-PCR amplification while Burkhardt et al. (14) compartmentalized the nested RT-PCR cocktail in a "tube-within-a-tube" device by using inexpensive materials such as a pipette tip and a microcentrifuge tube. Primers for the detection of hepatitis A virus by RT-PCR are summarized in Table 6.2.

ALTERNATIVE NUCLEIC ACID AMPLIFICATION METHODS

Nucleic acid sequence-based amplification (NASBA) is an amplification method that specifically detects RNA to the exclusion of DNA. The transcription-driven NASBA reaction is carried out at a single temperature (41°C) and theoretically amplifies the RNA target more than 10^{12}-fold within 90 min (34, 53). The final product of the amplification is single-stranded RNA, which can be readily detected by hybridization. The system utilizes three enzymes, (1) a reverse transcriptase (AMV-RT), (2) an RNase H, and (3) a T7 RNA polymerase, all of which act in a stepwise (sequential) manner with two oligonucleotide primers specific to the target (34, 53). One of the primers (P1) contains the T7 RNA

Table 6.1. Primers used in the detection of noroviruses in foods by RT-PCR

Primer	Viruses	Sequence (5'→3') (Polarity)	Location (bp)	Size (bp)	Food	Reference
Calman-1 Calman-2	NV (GII)	GCACACTGTGTTACACTTCC ACATTGGCTCTTGTCTGG	4193-4213 4997-5015	822	Clinical Stool	84
MJV12 RegA Nested p290	NV	TAY CAY TAT GAT GCH GAY TA CTC RTC ATC ICC ATA RAA IGA	4553-4572 4859-4879	326	Shellfish	78
Mp290 RevSR46 RevSR48-52	NV	GAT TAC TCC AAG TGG GAC TCC AC GAT TAT ACT SSM TGG GAY TCM AC CCA GTG GGC GAT GGA ATT CCA CCA RTG RTT TAT RCT GTT CAC	4568-4590 4786-4766	218	Shellfish	78
G1SKF G1SKR	NV GI	CTG CCC GAA TTT GTA AAT GA CCA ACC CAR CCA TTR TAC A	5342-5362	330	Fecal	62, 112
NI E3	NV GII	GAATTCCATCGCCCACTGGCT (+) ATCTCATCATCACCATA (−)	4756-4776 4869-4853	113	Shellfish	33, 37, 63
P290 P289	NVs and Sapoviruses	GATTACTCCAAGTGGGACTCCAC (+) TGACAATGTAATCATCACCATA (−)	4568-4590 4865-4886	319	Fruits and Vegetables	32
NVL1410U24 NVL1839L20	NV GI	(T/C)TT(T/C)TC(A/T/C)TT (T/C)TA(T/C)GG (G/T)GATGATGA GAA(G/C)CGCATCCA(G/A)CGGAACA	4489-4512	450	Shellfish	65
NVLII184U23 NVLII738L20	Norovirus GII	CA(G/A)(T/C)GGAACTCCA(T/C) (T/C)(G/A)CCCACTG (+) TGCGATCGCCCTCCCA(T/C)GTG (−)	5044-5063	574	Shellfish	65
Mon381	Norovirus GII	CCAGAATGTACAATGGTTATGC (+)	5362-5383	322	Deli Sandwich	22

Continued

Table 6.1. Primers used in the detection of noroviruses in foods by RT-PCR—*cont'd*

Primer	Viruses	Sequence (5' → 3') (Polarity)	Location (bp)	Size (bp)	Food	Reference
Mon383	Nested with 381	CAAGAGACTGTGAAGACATCATC (−)	5661–5683	223	Shellfish	95
Mon382		TGATAGAAATTGTTCCTAACATCAGG (−)	5559–5584			
NVp110	Noroviruses	AC(A/T/G)AT(C/T)TCATCATCACCAA (−)	4865–4884	398	Clams	59
NVp36		ATAAAAGTTGGCATGAACA (+)	4487–4501			87
JV12	Noroviruses	ATA CCA CTA TGA TGC AGA TTA (+)	4552–4572	326	Lettuce and	24
JV13		TCA TCA TCA CCA TAG AAA GAG (−)	4858–4878		Shellfish	

[1] GI: Genogroup I Noroviruses
[2] GII: Genogroup II Noroviruses

Table 6.2. Primers used in the detection of hepatitis A virus in foods by RT-PCR

Primer	Sequence (5' → 3') (Polarity)	Location (bp)	Size (bp)	Food	Reference
H1	GGAAATGTCTCAGGTACTTCTTTG (−)	2390–2414	247	Fruits/Vegetables	32
H2	GTTTTGCTCCTCTTTATCATGCTATG (+)	2167–2192		Mussels	101, 102
(HAVp3				Lettuce, Strawberry	12
HAVp4)				Shellfish	3, 4, 21, 70
(D, E)				Oysters	65, 69
				Delicatessen	91
HAV-R	CTCCAGAATCATCTCAAC (−)	2208–2226	192	Oyster	15, 16, 31, 51, 75
HAV-L	CAGCACATCAGAAAGGTGAG (+)	2035–2054		Lettuce, Hamburger	66, 87
				Mussel Tissue	82
2870	GACAGATTCTACATTTGGATTGGT (+)	2986–3004	534	Clams	59
3381	CCATTTCAAGAGTCCACACACT (−)	3381–3360			
BG7	CCGAAACTGGTTTCAGCTGAGG (−)	7125–7104	276	Produce and Shellfish,	35
BG8	CCTCTGGGTCTCCTTGTACAGC (+)	6850–6871		(Nested PCR)	
BG7a	CTGGTTTCAGCTGAGGYA (−)	7120–7102	264		
BG8a	GGTCTCCTTGTACAGCTT (+)	6856–6873			
2949	TATTTGTCTGTCACAGAACAATCAG (+)	2949–2973	267	Shellfish,	58
3192	AGGAGGTGGAAGCACTTCATTTGA (−)	3168–3145		(Nested PCR)	
dkA24	CTTCCTGAGCATACTTGAGTC (−)	3163–3182	200	Oysters	60
dkA25	CCAGAGCTCCATTGAACTC (+)	2986–3004			
Primer 1	CAGACTGTTGGGAGTGG (+)	762–778	385	Shellfish, (Nested PCR)	20
Primer 2	TTTATCTGAACTTGAAT (−)	1131–1147		Mussels, (Nested PCR)	19

Continued

Table 6.2. Primers used in the detection of hepatitis A virus in foods by RT-PCR—cont'd

Primer	Sequence (5' → 3') (Polarity)	Location (bp)	Size (bp)	Food	Reference
Primer 3	CAAGCACTTCTGTTTCCCGG (+)	780–797	329		
Primer 4	ATTGTCACCATAAGCAGCCA (−)	1092–1109			
P1	CAGGGGCATTTAGGTTT (−)	669–685	415	Produce (Nested PCR)	18
P2	CATATGTATGGTATCTCAACAA (+)	1063–1084			
P3	TGATAGGACTGCAGTGACT (−)	807–825	211		
P4	CCAATTTTGCAACTTCATG (+)	1000–1018			
HAV240	GGAGAGCCCTGGAAGAAAGA (−)	194–213	170	Bivalve Molluscs	85
HAV68	TCACCGCCGTTTGCCTAG (+)	43–60			
HAV1	TTGGAACGTCACCTTGCAGTG (+)	332–353	368	Shellfish (Nested PCR)	76
HAV2	CTGAGTACCTCAGAGGCAAAC (−)	680–700			
neHAV1	ATCTCTTTGATCTTCCACAAG (+)	371–391	290		
neHAV2	GAACAGTCCAGCTGTCAATGG (−)	641–661			

polymerase promoter sequence at 5′ terminal and the other primer (P2) can be designed with a generic sequence that facilitates probe capture for amplicon detection by liquid hybridization (described in the Confirmation section below).

NASBA assays have been developed for the detection of foodborne enteric viruses such as hepatitis A virus, rotavirus and noroviruses. For instance, Greene et al. (41) applied the NucliSens Basic Kit NASBA protocol for the detection of norovirus RNA in stool using primers targeting the RNA polymerase region of the viral genome. Jean et al. (54, 55, 56) developed a NASBA-based method to detect hepatitis A virus on artificially contaminated lettuce and blueberry samples and also for the detection of human rotavirus. More recently, Jean et al. (53) developed a multiplex NASBA method for the simultaneous detection of hepatitis A and noroviruses (GI and GII) from lettuce and sliced turkey (deli meats).

The NASBA method is isothermal, as rapid (if not faster than) as RT-PCR, and demonstrates detection limits equal to if not better than RT-PCR (41, 53, 56). However, NASBA technology has many of the same limitations as RT-PCR (e.g., contamination control, sample volume considerations, matrix-associated reaction inhibitors). Nonetheless, it remains an important alternative method for the detection of foodborne viruses.

CONFIRMATION

Since nonspecific products of amplification are a major issue when food and environmental samples are tested, it is critical to confirm that the nucleic acid amplification products obtained are specific to the target. Most often, the confirmation step also improves the sensitivity of the assay. The most common confirmatory tool is Southern hybridization using specific oligoprobes internal to the amplicon. These probes are usually enzyme labeled for colorimetric, luminescent or fluorescent endpoints. When RNA products (for NASBA) are tested, Northern hybridization with labeled internal oligonucleotide probes may be used (45, 46). An oligonucleotide array dot-blot format for the simultaneous confirmation of norovirus amplicons and strain genotyping has recently been reported, offering the promise of providing both detection and strain typing in a single test (107).

DNA enzyme immunoassay (DEIA) methods provide an alternative to Southern hybridization. In these assays, a capture probe is immobilized to a microtiter plate well and a labeled amplicon can then be detected directly, or alternatively, an unlabeled amplicon can be hybridized to a second labeled detector probe followed by detection after the addition of an enzyme-conjugate and appropriate substrate (sandwich assay). For colorimeteric, luminescent, or fluorescent endpoints, absorbance is read using a conventional microtiter plate spectrophotometer or fluorescent plate reader. The intensity of the signal obtained may be approximately proportional to the concentration of amplicon. The sensitivity of the microtiter plate assay is generally equal to or better than Southern hybridization and this approach has advantages including ease of interpretation, rapid (4 h) amplicon confirmation, and the potential for automation (54, 55, 90).

A liquid electrochemiluminescence (ECL) hybridization technology has been used commercially for the detection of NASBA amplicons. This technology utilizes two specific oligoprobes; a capture probe complementary to the sequence on primer P2, immobilized to streptavidin-labeled magnetic beads and a detector probe complexed to a ruthenium chelate. The hybridized magnetic particles are trapped on an electrode, and application of a voltage trigger to the electrode induces the ECL reaction such that the amount of emitted light is directly proportional to the amount of the amplicon. The signals are reported as ECL units by the NucliSens Reader and associated software. The NASBA-ECL system typically generates confirmed detection results in a day and has been used by Fox et al. (34) for the detection and confirmation of enterovirus from clinical samples and by Greene et al. (41) and Jean et al. (53) for the detection Norwalk virus in stool and food samples.

Other confirmation methods include specific "nested" PCR reactions (36, 43, 63, 99), which use a second pair of primers internal to the first amplicon sequence; and restriction endonuclease digestion of RT-PCR products (36, 43). Direct sequencing of the amplicon for the confirmation of RT-PCR products is another method of choice, and is frequently applied in the clinical realm, and more recently when amplicons are obtained from foods implicated in outbreaks (69, 91).

REAL-TIME DETECTION

Real-time detection refers to the simultaneous detection and confirmation of amplicon identity as the amplification reaction is progressing, thereby linking nucleic acid amplification with hybridization. There are currently five main chemistries used for real-time amplification and detection. One of the earliest and simplest approaches to real-time PCR, called DNA binding fluorophores, uses ethidium bromide or SYBR green I compounds that fluoresce when associated with double stranded DNA and exposed to a suitable wavelength of light. These methods tend to lack specificity but this has been addressed recently by coupling the assay with melting curve analysis. The 5' endonuclease assay (e.g., TaqMan oligoprobes, Applied Biosystems, Foster City, CA), adjacent linear probes (e.g., HybProbes, Roche Molecular Biochemicals, Germany) and hairpin oligoprobes (e.g. molecular beacons, Molecular Probes, Eugene, Oregon) have received considerable attention of late. Self-fluorescing amplicons (e.g., Sunrise primers, Amplifluor hairpin primers, Intergen Co., Purchase, NY) incorporated into the PCR product as the priming continues have a 3' end complimentary to the target strand and Scorpion primers have 5' end complementary to the target strand) (reviewed in 72). Recently, investigators have developed real-time PCR systems for the detection of a wide array of bacterial pathogens in foods. Prototype real-time RT-PCR amplification technologies have been developed for the detection of hepatitis A virus (17) and noroviruses (79) using the TaqMan format, and the norovirus detection uses the SYBR Green melting curve format (73). Beuret et al. (10) have used multiplex real-time PCR for the simultaneous detection of a panel of enteric viruses. Research efforts are currently underway to apply these methods to the detection of viruses in food matrices. Indeed, Myrmel et al. (78) recently reported the detection of viral contamination in shellfish using a commercial SYBRGreen PCR kit, while Narayanan et al (79) used their TaqMan assay to detect noroviruses in shellfish.

CONCLUSIONS

It should be clear from the preceding discussion that the current methodology for the detection of enteric viral contamination in foods is less than ideal and that research is necessary to improve these methods. Indeed, these protocols are applied infrequently and usually only in response to known or suspected foodborne disease outbreaks. The most important reasons for their limited use include: (1) the inability of molecular amplification methods to discriminate between infectious and inactivated virus, (2) the lack of widely accepted, collaboratively tested methods, (3) the requirement that most methods be product specific, meaning that universal approaches do not exist, and (4) the cost and need for highly trained personnel (83). When taken together, detection limits ranging from approximately 1–100 infectious units/g food have been obtained using various RT-PCR methods. The use of internal amplification standards to simultaneously evaluate RT-PCR inhibition and/or to provide a semiquantitative assay is also frequently done (5, 6, 67, 69, 89, 93, 103).

The failure to discriminate between infectious and inactivated virus is of critical importance because the inactivated forms of these pathogens pose no real public health threat. There is also a need to develop more universal sample extraction methods. For the most part, virus concentration from foods is likely to remain product dependent but research is needed to develop and refine the prototype methods into collaboratively tested protocols. Researchers continue to seek efficacious methods to concentrate the pathogens from the food matrix with the simultaneous removal of matrix-associated inhibitors. Even so, the methods will probably never be perfect and will always require a high degree of sample manipulation by the laboratory personnel (83). There is hope, however, that over time these rapid methods to detect human enteric viruses in foods may become more widely available to the food safety community.

ACKNOWLEDGMENTS

This effort was supported in part by a grant from the USDA National Research Initiative, Competitive Grants Program: Ensuring Food Safety, 2002-35201-11610. The use of trade names in this chapter does not imply endorsement by the North Carolina Agricultural Research Service nor criticism of similar ones not mentioned.

REFERENCES

1. Abd El Galil, K.H., M.A. El Sokkary, S.M. Kheira, A.M. Salazar, M.V. Yates, W. Chen, and A. Mulchandani. 2004. Combined immunomagnetic separation-molecular beacon-reverse-transcription-PCR assay for detection of hepatitis A virus from environmental samples. Appl. Environ. Microbiol. **70**:4371–4374.
2. Ando, T., S.S. Monroe, J.R. Gentsch, Q. Jin, D.C. Lewis, and R. I. Glass. 1995. Detection and differentiation of antigenically distinct small round-structured viruses (Norwalk-like viruses) by reverse transcription-PCR and Southern hybridization. J. Clin. Microbiol. **33**:64–71.
3. Arnal, C., V. Ferre-Aubineau, B. Besse, B. Mignotte, L. Schwartzbrod, and S. Billaudel. 1999. Comparison of 7 RNA extraction methods on stool and shellfish samples prior to hepatitis A virus amplification. J. Virol. Methods **77**:17–26.
4. Arnal, C., V. Ferré-Aubineau, B. Mignotte, B.M. Imbert-Marcille, and S. Billaudel. 1999. Quantification of hepatitis A virus in shellfish by competitive reverse transcription-PCR with coextraction of standard RNA. Appl. Environ. Microbiol. **65**:322–326.
5. Atmar, R.L., F.H. Neill, C.M. Woodley, R. Manger, G.S. Fout, W. Burkhardt, L. Leja, E.R. McGovern, F. LeGuyader, T.G. Metcalf, and M.K. Estes. 1996. Collaborative evaluation of a method for the detection of Norwalk virus in shellfish tissues by PCR. Appl. Environ. Microbiol. **62**:254–258.
6. Atmar, R.L., F.H. Neill, J.L. Romalde, F. Le Guyader, C.M. Woodley, T.G. Metcalf, and M.K. Estes. 1995. Detection of Norwalk virus and hepatitis A virus in shellfish tissues with the PCR. Appl. Environ. Microbiol. **61**:3014–3018.
7. Atmar, R.L., T.G. Metcalf, F.H. Neill, and M.K. Estes. 1993. Detection of enteric viruses in oysters by using polymerase chain reaction. Appl. Environ. Microbiol. **59**:631–635.
8. Barardi, C.R., M.H. Yip, K.R. Emslie, G. Vesey, S.R. Shanker, and K.L. Williams. 1999. Flow cytometry and RT-PCR for rotavirus detection in artificially seeded oyster meat. Int. J. Food Microbiol. **49**:9–18.
9. Berke, T., B. Golding, X. Jiang, W.D. Cubitt, M. Wolfaardt, A.W. Smith, and D.O. Matson. 1997. Phylogenetic analysis of the Caliciviruses. J Med Virol **52**:419–424.
10. Beuret, C. 2004. Simultaneous detection of enteric viruses by multiplex real-time RT-PCR. J. Virol. Methods **115**:1–8.
11. Beuret, C., A. Baumgartner, and J. Schluep. 2003. Virus-contaminated oysters: A three-month monitoring of oysters imported to Switzerland. Appl. Environ. Microbiol. **69**:2292–2297.
12. Bidawid, S., J.M. Farber, and S.A. Sattar. 2000. Rapid concentration of detection of hepatitis A virus form lettuce and strawberries. J. Virol. Methods **88**:175–185.
13. Boom, R., C.J.A. Sol, M.M.M. Salimans, C.L. Jansen, P.M.E. Wertheim-Van Dillen, and J.Van Der Noordaa. 1990. Rapid and simple method for purification of nucleic acids. J. Clin. Microbiol. **28**:495–503.
14. Burkhardt, W., G. Blackstone, D. Skilling, and A. Smith. 2002. Applied technique for increasing calicivirus detection in shellfish extracts. J. Appl. Microbiol. **3**:235–240.

15. Chung, H., L-A Jaykus, and M.D. Sobsey. 1996. Detection of human enteric viruses in oysters by in vivo and in vitro amplification of nucleic acids. Appl. Environ. Microbiol. **62**:3772–3778.

16. Coelho, C., A.P. Heinert, C.M.O. Simões, and C.R.M. Barardi. 2003. Hepatitis A virus detection in oysters (*Crassostrea gigas*) in Santa Caratina State, Brazil, by reverse transcription-polymerase chain reaction. J. Food Prot. **66**:507–511.

17. Costa-Mattioli, M, S. Monpoeho, E. Nicand, M-H Aleman, S. Billaudel, and V. Ferré. 2002. Quantification and duratin of viraemia during hepatitis A infection as determined by real-time RT-PCR. J. Viral Hepatitis **9**:101–106.

18. Croci, L., D. De Medici, G. Morace, A. Fiore, C. Scalfaro, F. Beneduce, and L. Toti. 2002. The survival of hepatitis A virus in fresh produce. Int. J. Food Microbiol. **73**:29–34.

19. Croci, L., D. De Medici, C. Scalfaro, A. Fiore, M. Divizia, D. Donia, A.M. Cosentino, P. Moretti, and G. Costantini. 2000. Deternination of enteroviruses, hepatitis A virus, bacteriophages and *Escherichia coli* in Adriatic Sea mussels. J. Appl. Microbiol. **88**:293–298.

20. Croci, L., D. DeMedici, G. Morace, A. Fiore, C. Scalfaro, F. Beneduce, and L. Toti. 1999. Detection of hepatitis A virus by nested reverse transcription-PCR. Int. J. Food Microbiol. **70**:67–71.

21. Cromeans T.L., O.V. Nainan, and H.S. Margolis. 1997. Detection of hepatitis A virus RNA in oyster meat. Appl. Environ. Microbiol. **63**:2460–2463.

22. Daniels, N.A., D.A. Bergmire-Sweat, K.J. Schwab, K.A. Hendricks, S. Reddy, S.M. Rowe, R.L. Fankhauser, S.S. Monroe, R.L. Atmar, R.I. Glass, and P. Mead. 2000. A foodborne outbreak of gastroenteritis associated with Norwalk-like viruses: First molecular traceback to deli sandwiches contaminated during preparation. J. Infect. Dis. **181**:1467–1470.

23. De Leon, R., S.M. Matsui, R.S. Baric, J.E. Herrmann, N.R. Blacklow, H.B. Greenberg, and M.D. Sobsey. 1992. Detection of Norwalk virus in stool specimens by reverse transcriptase-polymerase chain reaction and nonradioactive oligoprobes. J. Clin. Microbiol. **30**:3151–3157.

24. De Medici, D., Croci, L., E. Suffredini, and L. Toti. 2004. Reverse transcription-booster PCR for detection of noroviruses in shellfish. Appl. Environ. Microbiol. **70**:6329–6332.

25. De Medici, D., L. Croci, S. Di Pasquale, A. Fiore, and L. Toti. 2001. Detecting the presence of infectious hepatitis A virus in molluscs positive to RT-nested-PCR. Lett. Appl. Microbiol. **33**:362–366.

26. De Medici, D., F. Beneduce, A. Fiore, C. Scalfaro, and L. Croci. 1998. Application of reverse transcriptase-nested-PCR for detection of poliovirus in mussels. Int. J. Food Microbiol. **40**:51–56.

27. Demeke, T. and R.P. Adams. 1992. The effects of plant polysaccharides and buffer additives on PCR. Biotechniques **12**:332–333.

28. Deng M.Y., S.P. Day, and D.O. Cliver. 1994. Detection of hepatitis A virus in environmental samples by antigen-capture PCR. Appl. Environ. Microbiol. **60**:1927–1933.

29. Desenclos, J.C., K.C. Klontz, M.H. Wilder, O.V. Nainan, H.S. Margolis, and R.A. Gunn. 1991. A multistate outbreak of hepatitis A caused by the consumption of raw oysters. Am. J. Public Health **81**:1268–1272.

30. DiPinto, A., V.T. Forte, G.M. Tantillo, V. Terio, and C. Buonavoglia. 2003. Detection of hepatitis A virus in shellfish (*Mytilus galloprovincialis*) with RT-PCR. J. Food Prot. **66**:1681–1685.

31. Dix, A.B. and L.-A Jaykus. 1998. Virion concentration method for the detection of human enteric viruses in extracts of hard-shelled clams. J. Food Protect. **61**:458–465.
32. Dubois, E., C. Agier, O. Traoré, C. Hennechart, G. Merle, C. Crucière, and H. Laveran. 2002. Modified concentration method for the detection of enteric viruses on fruits and vegetables by reverse transcriptase polymerase chain reaction or cell culture. J. Food Prot. **65**:1962–1969.
33. Formiga-Cruz, M., G. Tofino-Quesada, S. Bofill-Mas, D. N. Lees, K. Henshilwood, A. K. Allard, A-C Conden-Hansson, B. E. Hernroth, A. Vantarakis, A. Tsibouxi, M. Papapetropoulou, M. D. Furones, and R. Girones. 2002. Distribution of human virus contamination in shellfish from different growing areas in Greece, Spain, Sweden, and the United Kingdom. Appl. Environ. Microbiol. **68**:5990–5998.
34. Fox, J.D., S. Han, A. Samuelson, Y. Zhang, M.L. Neale, and D. Westmoreland. 2002. Development and evaluation of nucleic acid sequence based amplification (NASBA) for diagnosis of enterovirus infections using the NucliSens® Basic Kit. J. Clin. Virol. **24**:117–130.
35. Goswami, B.B., M. Kulka, D. Ngo, P. Istafanos, and T.A. Cebula. 2002. A polymerase chain reaction-based method for the detection of hepatitis A virus in produce and shellfish. J. Food Prot. **65**:393–402.
36. Gouvea, V., N. Santos, M. Carmo-Timenetsky, and M.K. Estes. 1994. Identification of Norwalk virus in artificially seeded shellfish and selected foods. J. Virol Methods **48**:177–187.
37. Green, J., K. Henshilwood, C.I. Gallimore, D.W.G. Brown, and D.N. Lees. 1998. A nested reverse transciptase PCR assay for detection of small round-structured viruses in environmentally contaminated molluscan shellfish. Appl. Environ. Microbiol. **64**:858–863.
38. Green, J., C.I. Gallimore, J.P. Norcott, D. Lewis, and D.W.G. Brown. 1995. Broadly reactive reverse transcriptase polymerase chain reaction (RT-PCR) for the diagnosis of SRSV-associated gastroenteritidis. J. Med. Virol. **47**:392–398.
39. Green, S.M., P.R. Lambden, O. Caul, and I.N. Clarke. 1997. Capsid sequence diversity in small structured viruses from recent UK outbreaks of gastroenteritis. J. Med. Virol. **52**:14–19.
40. Green, S.M., P.R. Lambden, E.O. Caul, C.R. Ashley, and I.N. Clarke. 1995. Capsid diversity in small round-structured viruses: Molecular characterization of an antigenically distinct human enteric calicivirus. Virus Res. **37**:271–283.
41. Greene, S.R., C.L. Moe, L-A Jaykus, M. Cronin, L. Grosso, and P. van Aarle. 2003. Evaluation of the NucliSens® Basic Kit assay for the detection of Norwalk virus RNA in stool specimens. J. Virol. Methods **108**:123–131.
42. Groppe, J.C. and D.E. Morse. 1993. Isolation of full-length RNA templates for reverse transcription from tissues rich in RNase and proteoglycans. Anal. Biochem. **210**:337–343.
43. Hafliger, D., M. Gilgen, J. Luthy, and P. Hubner. 1997. Seminested RT-PCR systems for small round-structured viruses in faecal specimens. J. Virol. Methods **57**:195–201.
44. Hale, A.D., J. Green, and D.W.G. Brown. 1996. Comparison of 4 RNA extraction methods for the detection of small round structured viruses in faecal specimens. J. Virol. Methods **57**:195–201.
45. Hardy, M.E., S.F. Kramer, J.J. Treanor, and M.K. Estes. 1997. Human calicivirus genogroup II capsid sequence diversity revealed by analyses of the prototype Snow Mountain agent. Arch. Virol. **142**:197–202.

46. Honma, A., S. Nakata, K. Kinoshita-Numata, K. Kogawa, and S. Chiba. 2000. Evaluation of 9 sets of PCR primers in the RNA dependent RNA polymerase region for detection and differentiation of members of the family *Caliciviridae*, Norwalk virus and Sapporo virus. Microbiol. Immunol. **44**:411–419.

47. Iritani, N., Y. Seto, H. Kubo, K. Haruki, M. Ayata, and H. Ogura. 2002. Prevalence of "Norwalk-like virus" infections in outbreaks of acute nonbacterial gastroenteritis observed during the 1999–2000 season in Osaka city, Japan. J. Med. Virol. **66**:131–138.

48. Jansen, R.W., G. Siegl, and S.M. Lemon. 1990. Molecular epidemiology of human hepatitis A virus defined by antigen-capture polymerase chain reaction. Proc. Natl. Acad. Sci. USA **87**:2867–2871.

49. Jaykus, L., R. DeLeon, and G.A. Toranzos. 2001. Detection of bacteria, viruses and parasitic protozoa in shellfish, *In* C. J. Hurst (ed.), Manual of Environmental Microbiology, 2nd edn, pp. 264–276, ASM Press, American Society for Microbiology, Washington, DC.

50. Jaykus, L. 2000. Detection of human enteric viruses in foods. *In* S. Sattar (ed.), Foodborne Diseases Handbook, Vol. 2: Viruses, Parasites, Pathogens and HACCP, 2nd edn., pp. 137–163. Marcel Dekker, New York.

51. Jaykus, L.A., R. De Leon, and M.D. Sobsey. 1996. A virion concentration method for detection of human enteric viruses in oysters by PCR and oligoprobe hybridization. Appl. Environ. Microbiol. **62**:2074–80.

52. Jaykus L.A., R. De Leon, and M.D. Sobsey. 1995. Development of a molecular method for the detection of human enteric viruses in oysters. J. Food Protect. **58**:1357–1362.

53. Jean, J., D.H. D'Souza, and L. Jaykus. 2004. Multiplex nucleic acid sequence-based amplification (NASBA) for the simultaneous detection of enteric viruses in ready-to –eat food. Appl. Environ. Microbiol. **70**:6603–6610.

54. Jean, J., B. Blais, A. Darveau, and I. Fliss. 2002. Simultaneous detection and identification of hepatitis A virus and rotavirus by multiplex nucleic acid sequence-based amplification (NASBA) and microtiter plate hybridization system. J. Virol. Methods **105**:123–132.

55. Jean, J., B. Blais, A. Darveau, and I. Fliss. 2002. Rapid detection of human rotavirus using colorimetric nucleic acid sequence-based amplification (NASBA)-enzyme linked immunosorbent assay in sewage treatment effluent. FEMS Microbiol. Lett. **210**:143–147.

56. Jean, J., B. Blais, A. Darveau, and I. Fliss. 2001. Detection of hepatitis A virus by the nucleic acid sequence-based amplification (NASBA) technique and comparison with RT-PCR. Appl. Environ. Microbiol. **67**:5593–5600.

57. Jiang, X., P.W. Huang, W.M. Zhong, T. Farkas, D.W. Cubitt, and D.O. Matson. 1999. Design and evaluation of a primer pair that detects both Norwalk- and Sapporo-like caliciviruses by RT-PCR. J. Virol. Methods **83**:145–154.

58. Kingsley, D.H. and G.P. Richards. 2003. Persistence of hepatitis A virus in oysters. J. Food Protect. **66**:331–334.

59. Kingsley, D.H., G.K. Meade, and G.P. Richards. 2002. Detection of both hepatitis A virus and Norwalk-like virus in imported clams associated with food-borne illness. Appl. Environ. Microbiol. **68**:3914–3918.

60. Kingsley, D.H. and G.P. Richards. 2001. Rapid and efficient extraction method for reverse transcription-PCR detection of hepatitis A and Norwalk-like viruses in shellfish. Appl. Environ. Microbiol. **67**:4152–4157.

61. Kobayashi, S., K. Natori, N. Takeda, and K. Sakae. 2004. Immunomagnetic capture RT-PCR for detection of norovirus from foods implicated in a foodborne outbreak. Microbiol. Immunol. **48**:201–204.
62. Kojima, S., T. Kageyama, S. Fukushi, F.B. Hoshino, M. Shinohara, K. Uchida, K. Natori, N. Takeda, and K. Katayama. 2002. Genogroup-specific PCR primers for detection of Norwalk-like viruses. J. Virol. Methods **100**:107–114.
63. Lees, D. N., K. Henshilwood, J. Green, C. I. Gallimore, and D. W. Brown. 1995. Detection of small round structured viruses in shellfish by reverse-transcription-PCR. Appl. Environ. Microbiol. **61**:4418–4424.
64. Lees, D.N., K. Henshilwood, and W.J. Dore. 1994. Development of a method for detection of enteroviruses in shellfish by PCR with poliovirus as a model. Appl. Environ. Microbiol. **60**:2999–3005.
65. Legeay, O., Y. Caudrelier, C. Cordevant, L. Rigottier-Gois, and M. Lange. 2000. Simplified procedure for detection of enteric pathogenic viruses in shellfish by RT-PCR. J. Virol. Methods **90**:1–14.
66. Leggitt, P.R. and L.-A Jaykus. 2000. Detection methods for human enteric viruses in representative foods. J. Food Protect. **63**:1738–1744.
67. LeGuyader, F., L. Haugarreau, L. Miossec, E. Dubois, and M. Pommepuy. 2000. Three-y study to assess human enteric viruses in shellfish. Appl. Environ. Microbiol. **66**:3241–3248.
68. LeGuyader, F., M.K. Estes, M.E. Hardy, F.H. Neill, J. Green, D.W.G. Brown, and R.L. Atmar. 1996. Evaluation of a degenerated primer for the PCR detection of human caliciviurses. Arch. Virol. **141**:2225–2235.
69. LeGuyader, F., F.H. Neill, M.K. Estes, S.S. Monroe, T. Ando, R.L. Atmar. 1996. Detection and analysis of a small round-structured virus strain in oysters implicated in an outbreak of acute gastroenteritis. Appl. Environ. Microbiol. **62**:4268–4272.
70. Le Guyader, F., E. Dubois, D. Menard, and M. Pommepuy. 1994. Detection of hepatitis A virus, rotavirus, and enterovirus in naturally contaminated shellfish and sediment by reverse transcription-seminested PCR. Appl. Environ. Microbiol. **60**:3665–3671.
71. Lopez-Sabater, E.I., M.Y. Deng, and D.O. Cliver. 1997. Magnetic immunosepa-ration PCR assay (MIPA) for detection of hepatitis A virus (HAV) in American oyster (*Crassostrea virginica*). Lett. Appl. Microbiol. **24**:101–104.
72. Mackay, I.M., K.E. Arden, and A. Nitsche. 2002. Real-time PCR in virology. Nucl. Acids Res. **30**:1292–1305.
73. Miller, I., R. Gunson, and W.F. Carman. 2002. Norwalk like virus by light cycler PCR. J. Clin. Virol. **25**:231–232.
74. Moe, C., J. Gentsch, T. Ando, G. Grohmann, S.S. Monroe, X. Jiang, J. Wang, M.K. Estes, Y. Seto, C. Humphrey, S. Stine, and R.I. Glass. 1994. Application of PCR to detect Norwalk virus in fecal specimens from outbreaks of gastroenteritidis. J. Clin. Microbiol. **31**:2866–2872.
75. Mullendore, J.L., M.D. Sobsey, and Y.S. C. Sheih. 2001. Improved method for the recovery of hepatitis A virus from oysters. J. Virol. Methods **94**:25–35.
76. Muniain-Mujika, I.,M. Calvo, F. Lucena, and R. Girones. 2003. Comparative analy-sis of viral pathogens and potential indicators in shellfish. Int. J. Food Microbiol. **83**:75–85.
77. Muniain-Mujika, I., R. Girones, and F. Lucena. 2000. Viral contamination of shell-fish: Evaluation of methods and analysis of bacteriophages and human viruses. J. Virol. Methods **89**:109–118.

78. Myrmel, M., E.M. Berg, E. Rimstad, and B. Grinde. 2004. Detection of enteric viruses in shellfish from the Norwegian coast. Appl. Environ. Microbiol. 70:2678–2684.

79. Narayanan, J., J. Lowther, K. Henshilwood, D.N. Lees, V.R. Hill, and J. Vinje. 2005. Rapid and sensitive detection on noroviruses using Taq-Man-based one-step RT-PCR assays and application to clinical and shellfish samples. Appl. Environ. Microbiol. (In Press).

80. Noel, J.S., T. Ando, J.P. Leite, K.Y. Green, K.E. Dingle, M.K. Estes, Y. Seto, S.S. Monroe, and R.I. Glass. 1997. Correlation of patient immune responses with genetically characterized small round-structured viruses involved in outbreaks of nonbacterial acute gastroenteritidis in the United States, 1990–1995. J. Med. Virol. 53:372–383.

81. Ratcliff, R.M., J.C. Doherty, and G.D. Higgins. 2002. Sensitive detection of RNA viruses associated with gastroenteritis by a hanging-drop single-tube nested reverse transcription-PCR method. J. Clin. Microbiol. 40:4091–4099.

82. Ribao, C., I. Torrado, M.L. Vilarino, and J.L. Romalde. 2004. Assessment of different commercial RNA-extraction and RT-PCR kits for detection of hepatitis A virus in mussel tissues. J. Virol. Methods 115:177–182.

83. Richards, G. P. 1999. Limitations of molecular biological techniques for assessing the virological safety of foods. J. Food Protect. 62:691–697.

84. Rohayem, J., S. Berger, T. Juretzek, O. Herchenroder, M. Mogel, M. Poppe, J. Henker, A. Rethwilm. 2004. A simple and rapid single-step multiplex RT-PCR to detect Norovirus, Astrovirus and Adenovirus in clinical stool samples. J. Virol. Methods 118:49–59.

85. Romalde, J.L., E. Area, G. Sanchez, C. Ribao, I. Torrado, X. Abad, R.M. Pinto, J.L. Barja, and A. Bosch. 2002. Prevalence of enterovirus and hepatitis A virus in bivalve molluscs from Galicia (NW Spain): Inadequacy of the EU standards of microbiological quality. Int J Food Microbiol 74:119–130.

86. Rossen, L., P. Norskov, K. Holmstrom, and O.F. Rasmussen. 1992. Inhibition of PCR by components of food samples, microbial diagnostic assays and DNA extraction solutions. Int. J. Food Microbiol. 17:37–45.

87. Sair, A.I., D. D'Souza, C.L. Moe, and L-A Jaykus. 2002. Improved detection of human enteric viruses in foods by RT-PCR. J. Virol. Methods 100:57–69.

88. Sair, A.I., D.H.D'Souza, and L. Jaykus. 2002. Human enteric viruses as causes of foodborne disease. Comp. Rev. Food Sci. Safety 1:73–89.

89. Schvoerer, E., M. Ventura, O. Dubos, G. Cazaux, R. Serceau, N. Gournier, V. Dubois, P. Caminade, H.J. Fleury, and M.E. Lafon. 2001. Qualitative and quantitative molecular detection of enteroviruses in water from bathing areas and from a sewage treatment plant. Res. Microbiol. 152:179–186.

90. Schwab, K.J., F.H. Neill, F. Le Guyader, M.K. Estes, and R.L. Atmar. 2001. Development of a reverse transcription-PCR-DNA enzyme immunoassay for detection of "Norwalk-like" viruses and hepatitis A virus in stool and shellfish. Appl. Environ. Microbiol. 67:742–749.

91. Schwab, K.J., F.H. Neill, R.L. Fankhauser, N.A. Daniels, S.S. Monroe, D.A. Bergmire-Sweat, M.K. Estes, and R.L. Atmar. 2000. Development of methods to detect "Norwalk-like Viruses" (NLVs) and hepatitis A virus in delicatessen foods: Application to a foodborne NLV outbreak. Appl. Environ. Microbiol. 66:213–218.

92. Schwab, K.J., F.H. Neill, M.K. Estes, T.G. Metcalf, and R.L. Atmar. 1998. Distribution of Norwalk virus within shellfish following bioaccumulation and subsequent depuration by detection using RT-PCR. J. Food Prot. 61:1674–1680.

93. Schwab, K.J., M.K. Estes, F.H. Neill, and R.L. Atmar. 1997. Use of heat release and an internal RNA standard control in reverse transcription-PCR detection of Norwalk virus from stool samples. J. Clin. Microbiol. **35**:511–514.

94. Schwab, K.J., R. De Leon, and M.D. Sobsey. 1996. Immunoaffinity concentration and purification of waterborne enteric viruses for detection by reverse transcriptase PCR. Appl. Environ. Microbiol. **62**:2086–2094.

95. Shieh, Y.C., R.S. Baric, J.W. Woods, and K.R. Calci. 2003. Molecular surveillance of enterovirus and Norwalk-like virus in oysters relocated to a municipal-sewage-impacted gulf estuary. Appl. Environ. Microbiol. **69**:7130–7136.

96. Shieh, Y.C., S.S. Monroe, R.L. Frankhauser, G.W. Langlois, W. Burkhardt, III, and R.S. Baric. 2000. Detection of Norwalk-like virus in shellfish implicated in illness. J. Infect. Dis. **181**(S2):S360–366.

97. Shieh, Y.C., K.R. Calci, and R. S. Baric. 1999. A method to detect low levels of enteric viruses in contaminated oysters. Appl. Environ. Microbiol. **65**:4709–4714.

98. Straub, T.M., I.L. Pepper, and C.P. Gerba. 1994. Detection of naturally occurring enteroviruses and hepatitis A in undigested and anaerobically digested sludge using the polymerase chain reaction. Can. J. Microbiol. **40**:884–888.

99. Sugieda, M., K. Nakajima, and S. Nakajima. 1996. Outbreaks of Norwalk-like virus-associated gastroenteritis traced to shellfish: Coexistence of two genotypes in one specimen. Epidemiol. Infect. **116**:339–346.

100. Sunen, E. and M.D. Sobsey. 1999. Recovery and detection of enterovirus, hepatitis A virus and Norwalk virus in hardshell clams (*Mercenaria mercenaria*) by RT-PCR methods. J. Virol. Methods **77**:179–187.

101. Svensson, L. 2000. Diagnosis of foodborne viral infections in patients. Int. J. Food Microbiol. **59**:117–126.

102. Traore, O., C. Arnal, B. Mignotte, A. Maul, H. Laveran, S. Billaudel, and L. Schwartzbrod. 1998. Reverse transcriptase PCR detection of astrovirus, hepatitis A virus, and poliovirus in experimentally contaminated mussels: Comparison of several extraction and concentration methods. Appl. Environ. Microbiol. **64**:3118–3122.

103. Tsai, Y.L. and S.L. Parker. 1998. Quantification of poliovirus in seawater and sewage by competitive reverse transcriptase—Polymerase chain reaction. Can. J. Microbiol. **44**:35–41.

104. Tsai, Y-L., B. Tran, and C.J. Palmer. 1995. Analysis of viral RNA persistence in seawater by reverse-transcriptase-PCR. Appl. Environ. Microbiol. **61**:363–366.

105. Vinje, J., R.A. Hamidjaja, and M.D. Sobsey. 2004. Development and application of a capsid VP1 (region D) based reverse transcription PCR assay for genotyping of genogroup I and II noroviruses. J. Virol. Methods **116**:109–117.

106. Vinjé, J., H. Vennema, L. Maunula, L., C-H van Bonsdorff, M. Hoehne, E. Schreier, A. Richards, J. Green, D. Brown, S.S. Beard, S.S. Monroe, E. de Bruin, L. Svensson, and M P.G. Koopmans. 2003. International collaborative study to compare reverse transcriptase PCR assays for detection and genotyping of noroviruses. J. Clin. Microbiol. **41**:1423–1433.

107. Vinjé, J. and M.P. Koopmans. 2000. Simultaneous detection and genotyping of "Norwalk-like viruses" by oligonucleotide array in a reverse line blot hybridization format. J. Clin. Microbiol. **38**:2595–2601.

108. Wang, J., X. Jiang, H.P. Madore, J. Gray, U. Desselberger, T. Ando, Y. Seto, I. Oishi, J.F. Lew, K.Y. Green, and M.K. Estes. 1994. Sequence diversity of small, round-structured viruses in Norwalk virus group. J. Virol. **68**:5982–5990.

109. Wiedbrauk, D.L. and A.M. Devron. 1995. Nucleic acid detection methods. *In* D.L. Wiedbrauk and D.H. Farkas (eds.), Molecular Methods for Virus Detection, pp. 1–24. Academic Press Inc., San Diego, CA.

110. Wilde, J., J. Eiden, and R. Yolken. 1990. Removal of inhibitory substances from human fecal specimens for detection of group A rotaviruses by reverse transcriptase and polymerase chain reactions. J. Clin. Microbiol. **28**:1300–1307.
111. Wilson, J. G. 1997. Inhibition and facilitation of nucleic acid amplification. Appl. Environ. Microbiol. **63**:3741–3751.
112. Wolfaardt, M., M.B. Taylor, H.F. Booysen, L. Engelbrecht, W.O. Grabow, and X. Jiang. 1997. Incidence of human calicivirus and rotavirus infection in patients with gastroenteritis in South Africa. J. Med. Virol. **51**:290–296.
113. Yan, H., F. Yagyu, S. Okitsu, O. Nishio, and H. Ushijima. 2003. Detection of norovirus (GI, GII), sapovirus and astrovirus in fecal samples using reverse transcription single-round multiplex PCR. J. Virol. Methods **114**:37–44.

CHAPTER 7

Molecular Tools for the Identification of Foodborne Parasites

Ynes Ortega*, Ph.D.

Center for Food Safety, College of Agriculture and Environmental Sciences, The University of Georgia, Griffin, GA 30223

Introduction
DNA Extraction Procedures
Protozoal Infections
Cryptosporidium parvum
    Parasite Description and Identification
    Molecular Detection
Cyclospora cayetanensis
    Parasite Description and Identification
    Molecular Detection
Giardia intestinalis
    Parasite Description and Identification
    Molecular Detection
Toxoplasma gondii
    Parasite Description and Identification
    Molecular Detection
Microsporidia
    Parasite Description and Identification
    Molecular Detection
    Helminth Infections
    Viability Assays
    Conclusions
    References

INTRODUCTION

Parasites have long been associated with food and waterborne outbreaks. Although parasites have been consistently reported in developing and endemic countries, the number of parasites present in the food supply of Americans has multiplied by more than a factor of 8 during the past 15 years. This is partly due to the increase of international travel and population migration. Rapid and refrigerated food transportation from foreign countries facilitates consumer contact with emerging parasites. Cultural habits have also changed towards the

*Corresponding author. Phone: (770) 233-5587; e-mail: ortega@uga.edu

consumption of (raw or undercooked) fresh produce. These conditions have increased the probability that parasites infect naive populations and cause gastrointestinal illness (91).

Most parasites are obligate intracellular organisms. In contrast with bacteria, parasites are inert and do not multiply in the environment. Any isolation and detection procedures are crucial because an enrichment process for parasites is not available. Molecular assays overcome these difficulties and specific limitations per organism will be discussed.

Based on their morphological attributes, parasites are classified into two groups: protozoa and helminths. The protozoa are single-celled organisms and the helminths are metazoans with a rudimentary digestive and reproductive tract. The helminths are grouped as the nematode or roundworms, the cestoda or tapeworm, and the trematoda or flukes. The primary objectives in developing diagnostic molecular assays have been focused on protozoa because of the limitations of conventional parasitological methods to identify them in foods and the environment.

DNA EXTRACTION PROCEDURES

Many protocols have been described for the isolation of parasite DNA. Some are protocols prepared at individual laboratories. The trend is now to use DNA extraction kits, as this will reduce the possible variables among procedures and laboratories. The recovery efficiency of the extraction procedures, particularly with parasites, is important because of the potentially low number of parasites in environmental samples and food matrices. The sensitivity of the PCR or any molecular assay is dependent on the DNA extraction methodology. Enzyme digestion of the protozoal oocysts has been done using proteinase K in lysis buffer (10 mg proteinase K/ml, 120 mM $NaCl_2$, 10 mM Tris and 0.1% SDS) followed by phenol chloroform-isoamyl alcohol (25:24:1) separation, and DNA precipitation using salts such as 0.3 M sodium acetate with 10 µg glycogen, a DNA carrier, and 2 volumes of 100% ethanol. This method extracts DNA efficiently from parasites, but the most important step in this process is to break open the oocysts to release the protozoal DNA. Not all protozoal cysts or oocysts will be effectively digested; therefore other means of oocyst rupture have been described. One of the commonly used methods for cyst/oocyst breakage is the freeze/thaw method with cycles of freeze/thaw that vary from 3 to 12 cycles. The freezing is done in dry ice/ethanol slurry and thawed at 55°C. This process may change accordingly to the parasite and the laboratory (114). Oocysts can also be disrupted using a bath sonicator (101), keeping under consideration the possible denaturation of the DNA. Based on forensic studies, an extraction free, filter based preparation of DNA has been used and can successfully rupture the parasite cysts or oocysts. This method uses an FTA filter, which is impregnated with denaturants, chelating agents, and free radical traps. The sample is placed on the filter and cut with a hole puncher. The membrane is then rinsed and used directly for PCR amplification (92). The use of this methodology needs to be evaluated for the processing and evaluation of large sample sizes and the potential for cross-contamination.

PROTOZOAL INFECTIONS

Protozoan parasites relevant to public health and associated with foodborne infections include the ciliates (i.e., *Balantidium coli*), amoeba (i.e., *Entamoeba histolytica*), flagellates (i.e., *Giardia lamblia*), and coccidia (i.e., *Toxoplasma gondii*, *Cryptosporidium parvum* and *Cyclospora cayetanensis*). Taxonomical placement of some protozoan pathogens has been changing as we learn more about the genetic makeup of these organisms. Such is the case with the Microsporidia, which are now considered to be more closely related to fungi than to the protozoa. *Cryptosporidium*, originally from the phylum *Apicomplexa*, has more characteristics associated to the gregarines than to coccidia. Its taxonomical classification is uncertain and as more evidence is published, these parasites may be reclassified to more appropriate groups.

CRYPTOSPORIDIUM PARVUM

Parasite Description and Identification. *Cryptosporidium* sp. was first recognized by Tyzzer in 1907, from the stomach of a mouse (78). It is currently identified as *C. muris*. Subsequently, other species of *Cryptosporidium* were described and renamed. *Cryptosporidium* species that were morphologically similar were named *Cryptosporidium parvum*, but recent molecular analysis have lead to reclassification into several species: *C. hominis* (*C. parvum* genotype 1 or the human genotype) in humans, *C. andersoni* (*C. muris*-like or *C. muris* bovine genotype) and *C. bovis* (*Cryptosporidium* bovine genotype B) in calves and adult cattle, *C. canis*, (*C. parvum* dog genotype) in dogs, and *C. suis* (*Cryptosporidium* pig genotype I) in pigs. To date, there are 15 established *Cryptosporidium* species in fish, reptiles, birds, and mammals and 8 have been reported in humans (*C. hominis*, *C. bovis*, *C. canis*, *C. felis*, *C. meleagridis*, *C. muris*, *C. suis*, and *Cryptosporidium* cervine genotype). *Cryptosporidium* has been associated with gastrointestinal illness in humans and it is acquired by ingestion of fecally contaminated water or foods. It can also be acquired via person to person.

The life cycle of *Cryptosporidium* starts when mature and infectious oocysts are excreted in the feces of an infected host. The oocysts are ingested along with contaminated water and/or foods and excyst in the gastrointestinal tract. Sporozoites are released and infect the epithelial cells of the small intestine, particularly the ileum. Parasites multiply asexually, producing type I and II meronts containing 8 and 4 merozoites, respectively. Asexual multiplication can continue or differentiate to produce the sexual stages of the parasites. Microgametocytes (male) fertilize the macrogametocyte (female) producing the zygote, which in turn becomes the oocyst. If a thin-walled oocyst is formed, the life cycle can initiate again. If a thick-walled oocyst (which is environmentally resistant) is formed, it is excreted in the feces (78).

Individuals at risk of acquiring cryptosporidiosis include children in daycare centers, individuals caring for animals, the elderly, immunocompromised individuals, and travelers. Since *Cryptosporidium* is highly resistant to common disinfectants, including chlorine, water parks and fountains have been implicated in numerous outbreaks.

A large waterborne outbreak occurred in 1993 in Milwaukee (76, 146), where more than 400,000 people suffered gastrointestinal illness. Initially, *C. parvum* was considered to be the agent responsible for this outbreak, however, molecular analysis of clinical specimens and water samples demonstrated that it was actually *C. hominis*. Contamination occurred when tap water was contaminated with sewer water back-flow (76, 146).

Two main target antigens of 15-17 and 23 kDa molecular weights have been used for detection of the humoral immune response of individuals with cryptosporidiosis (102). These antigens have been used in ELISA, western blot, or multiplex bead assay for the detection of *Cryptosporidium* antibodies in sera and oral fluids (85, 99, 148).

Stool specimens are the most common clinical samples examined for diagnosis of *Cryptosporidium*. Oocysts are usually concentrated using ethyl acetate concentration methodologies (139). The sample is then examined using microscopy, immunoassays, and molecular techniques. The modified Ziehl-Neelsen acid-fast stain and modified Kinyoun's acid-fast stain are more commonly used in the microscopic identification of *Cryptosporidium* (87, 103).

Immunofluorescence assays (IFA) are more sensitive and specific than the modified acid fast stains and are now being used more frequently in clinical laboratories. They are the gold standard when examining new diagnostic assays (9, 40), but cannot differentiate among the *Cryptosporidium* species (46). Some of these commercial kits include the Merifluor *Cryptosporidium/Giardia* kit (Meridian Bioscience; Cincinatti, OH), *Giardia*/Crypto IF kit (TechLab; Blacksburg, VA), Monofluo *Cryptosporidium* kit (Sanofi Diagnostics Pasteur), Crypto/*Giardia* Cel kit (TCS Biosciences; Buckingham, UK), and Aqua-Glo G/C kit (Waterborne; New Orleans, LA). Commercial antigen-capture-based enzyme immunoassays (EIA) available are the Alexon-Trend ProSpecT *Cryptosporidium* Microplate Assay (Alexon-Trend-Seradyn; Ramsey, MN) and Meridian Premier *Cryptosporidium* kit. These assays may not react to *Cryptosporidium* species that are genetically distant from *C. parvum*, such as *C. muris*, *C. andersoni*, *C. serpentis* and *C. baileyi* (46). Lateral flow immunochromatographic assays have also been commercialized for use with stool samples (60)

Parasites can be recovered from foods by washing the samples in 0.025M phosphate buffered saline, pH 7.25 (95). Detergents (1% sodium dodecyl sulfate and 0.1% Tween 80, or the membrane filter elution buffer from EPA method 1623) and sonication (3-10 min) are also used to facilitate the elution of parasites from the food matrices (13, 105). *Cryptosporidium* can then be concentrated by centrifugation and examined by immunofluorescence staining (13, 95). A sucrose flotation step may be included producing a cleaner sample but at the cost of losing parasites. Moderate recovery rates of 18.2-25.2% were reported for a variety of fresh produce. Immunomagnetic separation has been included in the *Cryptosporidium* recovery procedures from lettuce, Chinese leaves, and strawberries to 42% for *Cryptosporidium* and 67% for *Giardia* (105, 106). *Cryptosporidium* oocysts can be detected by IFA in shellfish gills, gastric glands, and hemocytes from the hemolymph (32, 49).

Identification of *Cryptosporidium* oocysts in water samples is achieved by using IFA after concentration processes (EPA ICR method, EPA method1622/1623, United Kingdom SCA method, and United Kingdom regulatory method) (72). Oocysts are recovered by filtration of 10–100 L or more water, concentrated and stained with FITC-labeled *Cryptosporidium* antibodies. The ICR or SCA methods use nominal 1 µm 10″ cartridge filters and washes concentrated by flotation (using Percoll, sucrose, or potassium citrate). The EPA method 1622 and the United Kingdom regulatory method use capsule filters for filtration followed by immunomagnetic separation (IMS). To determine oocyst viability, 4′, 6-diamidoino-2-phenylindole (DAPI) vital dye is used. The recovery rates of the EPA method 1622/1623 for *Cryptosporidium* oocysts are between 10 and 75% for surface water (67, 115, 137). One of the limitations of these procedures is the cross-reactivity of the monoclonal antibodies used in the IMS and IFA kits. Dinoflagellates (120) and algae (111) may not provide accurate detection and quantification of *Cryptosporidium* oocysts, and may require confirmation by differential interference contrast microscopy.

Surface water samples may contain *Cryptosporidium* oocysts, which are a nonpathogenic species or genotypes for humans. Therefore, identification of oocysts to the species/genotype level is of significant public health importance. The PCR has become a useful tool for determining the genotype *Cryptosporidium* oocysts found in shellfish (32, 44, 45).

Analysis of environmental samples presents several limitations to these assays. The number of parasites in foods is usually small. Other structures morphologically similar may nonspecifically react with antibody-based assays for *Cryptosporidium* and make them less reliable. In addition, the presence of *Cryptosporidium* oocysts, which are not infectious to humans, may be present and could be confused with those of public health relevance. When examining environmental samples (soil, water, or foods), density gradients or antibody-based concentration procedures have been described (i.e., IMS concentration). Concentrates are then used for direct-fluorescent antibody (DFA) or for PCR detection.

Molecular Detection. One of the advantages of using molecular assays on environmental samples is the capability of identifying low numbers of parasites and being able to determine their species and genotype. This allows for parasite fingerprinting in waterborne and foodborne outbreaks, particularly when determining if the parasite is anthroponotic or zoonotic. It also aids in determining the risk factors associated with transmission of cryptosporidiosis in a particular setting (4, 5, 43, 66). Subtyping tools have been useful in the investigation of foodborne and waterborne outbreaks of cryptosporidiosis (43, 71, 124).

Earlier PCR methods (22, 65, 140) have been used only for the identification of *Cryptosporidium* spp. Several PCR-Restriction Fragment Length Polymorphism (RFLP)-based genotyping tools have been developed for the detection and differentiation of *Cryptosporidium* at the species level (6, 62, 70, 74, 86, 120, 146). Most of these techniques are based on the SSU rRNA gene. However, some of the SSU rRNA-based techniques (62, 70) used conserved sequences of eukaryotic organisms, which amplify DNA from organisms other than *Cryptosporidium* (127). Nucleotide sequencing-based approaches have also been developed for the differentiation of various *Cryptosporidium* spp.(82, 84, 125, 126, 136). These techniques use long PCR amplicons, and some amplify other Apicomplexan parasites and dinoflagellates. Because of this lack of specificity and sensitivity they cannot be used for diagnostic purposes (125, 126, 136).

Other genotyping techniques are used specifically to differentiate *C. parvum* from *C. hominis* (15, 83, 97, 118, 141). Out of ten commonly used genotyping tools for *Cryptosporidium* species/genotypes, only the SSU rRNA-based PCR tools can detect all seven *Cryptosporidium* species/genotypes (57).

Microsatellite analysis has been used to characterize diversity between *C. parvum* or *C. hominis* and their subtypes (19, 20, 35, 142). High-sequence polymorphism in the gene of 60 kDa glycoprotein precursor has also been used for subtype analysis (98, 124). Other subtyping tools include sequence analysis of HSP70 (98, 124), heteroduplex analysis and nucleotide sequencing of the double-stranded RNA (71, 146), and single-strand conformation polymorphism (SSCP)-based analysis of the internal transcribed spacer (ITS-2) (41–42).

The SSU rRNA-based nested PCR-RFLP method has been successfully used in conjunction with IMS in the detection and differentiation of *Cryptosporidium* oocysts present in storm water, raw surface water, and wastewater (147, 148).

Two SSU rRNA-based PCR-sequencing tools and one other SSU-based PCR-RFLP tool can differentiate *Cryptosporidium* oocysts in surface and wastewater samples (86, 136), suggesting that humans, farm animals, and wildlife contribute to *Cryptosporidium* oocyst contamination in water. Other genes have also been used for genotyping of *Cryptosporidium*. HSP70 and TRAP-C2-genes have limited use as they do not amplify DNA of *Cryptosporidium* species distant from *C. parvum*(57).

A few PCR related techniques have also been used to quantify and determine viability of *Cryptosporidium* oocysts. Excystation followed by DNA extraction and PCR has been developed to detect viable *C. parvum* oocysts (36, 134). Tissue culture and PCR (26, 67, 108) or reverse transcription-PCR (RT-PCR) (108, 109) has been used to detect viable *Cryptosporidium* oocysts. RT-PCR techniques have been described for the detection of viable oocysts (48, 56, 143), but it may overestimate the viability of oocysts (37). A new integrated detection assay combining capture of double-stranded RNA with probe-coated beads, RT-PCR, and lateral flow chromatography has also been developed, which should shorten detection time (64).

Other molecular tools, such as fluorescence *in situ* hybridization (FISH), or colorimetric *in situ* hybridization of probes to the SSU rRNA have been used in the detection or viability evaluation of *C. parvum* oocysts (72, 116). Nucleic acid sequence-based amplification (NASBA) has been used in the detection of viable *C. parvum* oocysts (10). More recently, a biosensor technique for the detection of viable *C. parvum* oocysts has also been described (11), and a microarray technique based on HSP70 sequence polymorphism has been developed to differentiate *Cryptosporidium* genotypes (119).

CYCLOSPORA CAYETANENSIS

Parasite Description and Identification. *Cyclospora cayetanensis* was initially identified as a cyanobacteria-like organism. The first reports of gastrointestinal infections in humans go back to 1988, when it was described as a blue-green algae or a large *Cryptosporidium*. It was not until 1992 when Ortega et al. reported a complete description of this parasite as belonging to the coccidian (93, 96). *Cyclospora* has been identified in insectivores, snakes, and rodents. In 1997, three species were identified in nonhuman primates (30, 73). These are morphologically similar to *C. cayetanensis*, but examination of the 18S rDNA demonstrated that they are phylogenetically different. A challenge when working with *Cyclospora* is that is there is no animal model suitable to propagate it in laboratory conditions. The same limitation applies to the nonhuman primate *Cyclospora* species.

Cyclosporiasis is characterized by prolonged watery diarrhea. Nondifferentiated oocysts are excreted in the feces of the infected individual into the environment. Oocysts differentiate after 7-15 days in the environment, becoming fully sporulated and infectious. When a susceptible individual ingests oocysts from contaminated water or foods, the oocyst will excyst and release the sporocysts, which also will undergo excystation (96). Each contains two sporozoites that will infect the epithelial cells of the small intestine (94). *Cyclospora* preferentially colonizes the ileum; however, there are few reports suggesting extraintestinal colonization such as the biliary and respiratory tracts.

Even though there have been few reports of *Cyclospora* associated with drinking or swimming in contaminated water, most of the reported outbreaks in the developed world have been associated with contaminated fresh produce and berries. Lettuce, basil, and raspberries have been the most frequently implicated products (50). In 2004, the first documented outbreak of cyclosporiasis was linked to Guatemalan snow peas (113). Some patients reported consumption of untreated water or reconstituted milk.

Diagnosis of *Cyclospora* in clinical specimens is performed using a modified acid fast stain, or by direct microscopical examination. A procedure to isolate *Cyclospora* oocysts from produce was described by Robertson (107). For mushrooms, lettuce, and raspberries, recovery was 12%; while recovery from bean sprouts was 4% using lectin-coated paramagnetic beads to concentrate the parasite. Although the lectin-coated paramagnetic beads did not significantly improve recovery of the oocysts, it did produce a cleaner and smaller final volume for easier identification of the protozoan under the microscope.

Molecular Detection. The first PCR for *Cyclospora* was developed for clinical specimens. Fecal samples containing oocysts were disrupted using a bath sonicator at 120 W. The product of the nested PCR was a 304 bp amplicon (101). However, this PCR produced amplicons for *Eimeria*; a coccidian that infects a wide range of animals commonly found in environmental samples and is nonpathogenic to humans. This nested PCR was modified using the same primers but without the leader sequence. The template was prepared using 6 freeze/thaw cycles (2 min in liquid nitrogen and heating at 98°C). To address the issue of PCR inhibitors, Instagene Matrix (Bio-Rad; Hercules, CA) was added during the DNA extraction. Nonfat milk (50 mg/ml) was used in the PCR reaction to overcome the effect of inhibitory effects of the food matrix extracts, soil, and plant matrices. An RFLP PCR was also developed to differentiate between *Eimeria* and *Cyclospora* (58).

An extraction free filter based template preparation was evaluated using *Cyclospora* oocysts from fecal and food matrix samples. Pieces of the FTA filters (Whatman; Florham Park, NJ) were added directly to the PCR mixture after washing with 10 mM Tris (pH8.0) containing 0.1 mM EDTA and heat treatment at 56°C. A sensitivity of 10 to 30 *Cyclospora* oocyst could be detected in 100 g of fresh raspberries (92).

Quantitative real-time PCR was developed for the identification of Cyclospora, targeting a 83-bp region of the 18s rRNA gene. This analysis was based on one *Cyclospora* isolate. The sensitivity and specificity of the assay will need to be confirmed when examining a larger number of environmental samples (132).

Since other nonhuman primate *Cyclospora* also produced 294 bp amplicon with the SSU-rRNA nested PCR primers, a multiplex PCR was developed. Briefly, PCR is performed first using external, nested PCR primers described by Jinneman, followed with PCR using a series of specific primers that differentiate *Cyclospora* species and *Eimeria* species (90).

A restriction fragment length polymorphism (RFLP) assay was developed using the endonuclease *Mnl* I. RFLP DNA patterns are different between *Cyclospora* and *Eimeria* (58). To simplify the methodology for use at inspection sites, an oligo–ligation assay (OLA) was developed. The target amplicon was detected with an antibody-enzyme conjugate that could be read as a colorimetric assay (59).

Methodologies that could be used for fingerprinting analysis and genotype discrimination are not yet available. The intervening transcribed spacer 1(ITS1) was examined as a potential sequence that will allow discrimination among *Cyclospora* genotypes. When compared, the sequences of 5 isolates from a *Cyclospora* foodborne outbreak were identical. One of two Guatemalan isolates and 2 out of 4 Peruvian isolates were also identical (2). Thirty-six *Cyclospora* samples were examined using ITS1 specific primers. Sequence homology of 460–465 bp from various isolates varied between oocyst samples from 1 to 6.5% and between samples 0 and 5.7%. *Cryptosporidium* ITS is also highly variable (1.1–1.3% within and 0.8–1.6% between oocyst samples). The lack of animal models to obtain a clonal population of *Cyclospora* limits the possibility to determine if there are different ITS sequence types within one oocyst or if there are infections with various *Cyclospora* strains (88). If intraisolate variation occurs, as with many coccidian parasites, ITS1 may not be a suitable target for genotyping of *Cyclospora*.

GIARDIA INTESTINALIS

Parasite Description and Identification. *Giardia* was initially described by Leeuwenhoek in 1681 (1). *Giardia* infecting humans was renamed in the early 1990s: as *G. intestinalis, G. lamblia,* and *G. duodenalis.* The name *G. lamblia* was well-recognized in the 1970s, but encouraged by other investigators it was changed to *G. duodenalis* and *G. intestinalis* in the 1990s. Over 40 species names have been proposed on the basis of hosts of origin. Now, several species are recognized: *G. muris* from rodents, *G. agilis* from amphibians, *G. psittaci* from parakeets, *G. ardae* from herons, and *G. microti* from voles and muskrats. *Giardia* is the most common cause of waterborne outbreaks; however, there have also been reports of foodborne outbreaks. In developing countries, chronic giardiasis has been associated with long-term growth retardation. *Giardia* can be asymptomatic or cause intermittent or chronic diarrheal complaints. Symptomatic giardiasis is characterized by prolonged and intermittent diarrhea, anorexia, flatulency, weight loss, and malabsorption. Whether *Giardia lamblia* infects humans and other mammals, whether it is a single species, and if so, whether it is a zoonotic infection has been questioned. Foods implicated with giardiasis are fresh produce or foods contaminated by food handlers (1).

The infective stage of *Giardia* is the environmentally resistant cyst and the vegetative stage is the trophozoite. The trophozoite is pear shaped and has 8 flagella and a ventral adhesive disk. When a susceptible individual ingests contaminated water or foods containing the cysts, these cysts will excyst in the intestine aided by bile salts and gastric acids. The trophozoite will multiply asexually. *Giardia* does not invade tissues and propagation occurs on the epithelial surface. The trophozoites attach to the epithelial surface. Some of the trophozoites will encyst in the jejunum and are passed in the feces.

Giardia is the most commonly isolated gastrointestinal parasite worldwide and the U.S., (61) with large waterborne outbreaks having been reported. The cysts are environmentally resistant and survive long periods of time in water at cold temperatures. *Giardia* has been detected in 81% of raw water samples and 17% of filtered water samples (68). Children attending day care centers are at higher risk of acquiring Giardiasis. Waterborne Giardiasis has been associated with unfiltered water; recreational water, such as swimming pools; water fountains; and travel (78).

Molecular classification tools have been used with various *Giardia* isolates and have determined various assemblages or genotypes based on the sequence comparisons of the SS rRNA, triosephosphate isomerase, and glutamate dehydrogenase genes. Genotype A groups 1 and 2 have been isolated from humans and animals. Genotype B has also being isolated from humans and some animals. Assemblages C and D have been isolated from dogs, F from cats, G from rats, and E from cows, sheep alpaca, goat, and pigs. The significance of these assemblages in the human infections is being examined (1). In the Netherlands, patients with assemblage A isolates presented with intermittent diarrheal complaints, while assemblage B was present in individuals with persistent diarrheal complaints (51).

Diagnosis of *Giardia* in clinical samples is performed using bright field microscopy. Immunofluorescent assay and EIA are commercially available. Merifluor *Cryptosporidium*/*Giardia* kit (Meridian Bioscience), *Giardia*/Crypto IF kit (TechLab), Crypto/*Giardia* Cel kit (TCS Biosciences), and Aqua-Glo G/C kit (Waterborne) are some of the commercially available kits. Detection of *Giardia* cysts in environmental samples presents a significant challenge. Recovery of *Giardia* cysts from surface water to be used by drinking water treatment plants is achieved using the United States EPA Method 1623. Several of these methods have been adapted for detecting parasites in food matrices.

Molecular Detection. However, the low number of cysts typically present in these types of samples requires the use of molecular tools. In clinical specimens *Giardia* and *Cryptosporidium* were detected 22 times more often by PCR than by conventional microscopy (7).

Analysis of nucleotide sequences of glutamate dehydrogenase (GDH), elongation factor 1α (EF1α), SSU rRNA, and triosephosphate isomerase (TPI), and ADP-ribosylating factor genes can discriminate five to seven defined lineages and assemblages of *G. intestinalis* (80, 81, 121). Thus far, the TPI gene has the highest polymorphism in *G. intestinalis* at both intergenotype as well as intragenotype levels, and TPI genotyping has proven very useful in epidemiological investigations of human Giardiasis (121, 123).

Molecular characterization of *Giardia* species in wastewater has been used for community wide surveillance of human Giardiasis (130, 135). The distribution of the *Giardia* species in environmental samples correlates directly with human, agricultural, and wildlife activities. SSU rRNA-based PCR-RFLP (130) and beta-giardin-based PCR-RFLP methods were used to identify and differentiate between *Giardia* assemblages A and B from water samples.

The phylogenetic distance between *G. intestinalis* assemblages A and B is greater than typically used to differentiate two protozoan species (79, 81, 121, 131), suggesting that *G. intestinalis* may be a species complex. Phenotypic differences have also been observed. Assemblage B is more likely found in patients with persistent diarrhea, whereas intermittent diarrhea is mostly observed with assemblage A (51).

TOXOPLASMA GONDII

Parasite Description and Identification. *Toxoplasma* is a coccidian parasite that can cause severe complications for individuals with the infection. It can be asymptomatic or can cause abortion in humans if an acute infection develops during pregnancy. Healthy individuals may develop encephalitis or be asymptomatic. *Toxoplasma* can be found worldwide and serologically it can be identified in as high as 85% of the population in some European countries. *Toxoplasma* is responsible for 20.7% of foodborne deaths due to known infectious agents. Waterborne outbreaks in Canada and Brazil (12, 54) have been reported as well.

Infection occurs when water or foods are contaminated with cat feces containing *Toxoplasma* oocysts. The oocysts excyst and the sporozoites migrate and preferentially localize in muscle and the brain. The parasites will encyst and form cysts. These contain bradyzoites, which are slow multiplying parasites. Once they become active they are called tachyzoites and multiply quickly. The parasite can cross the placenta to infect the fetal tissues (29).

A large variety of animals can acquire toxoplasmosis, but only felines (domestic and wild) are the definitive hosts. When infected tissues are ingested, the parasites will be released from the tissues and develop to the asexual and sexual stages. Once fertilization occurs, the oocysts are formed and then excreted in the environment. The oocysts are highly resistant even to desiccation and survive on dry surfaces for weeks or even months. *Toxoplasma* causes fatal meningoencephalitis in a variety of marine mammals. Shellfish has been studied during the past several years as indicators of or vectors for transmission of protozoal agents (29). Shellfish can concentrate large volumes of water and it has been demonstrated experimentally that viable *Toxoplasma* oocysts can also be concentrated. *Toxoplasma* has been classified into three different lineages I, II, and III (53). *Toxoplasma* belonging to type I lineage are highly virulent in laboratory animals, whereas the type II and III lineages are nonvirulent. In humans, the type II lineage predominates among *Toxoplasma* associated with infections in AIDS and non-AIDS immunocompromised patients (75–80%), as well as congenital *Toxoplasma* infections (encephalitis, pneumonitis, or disseminated infections). It appears that the type I lineage is more prevalent in congenital *Toxoplasma* infections in Spain (38). Ocular toxoplasmosis is a common sequela of congenital toxoplasmosis, but can be dormant for years and emerge at adulthood; causing severe retinochoriditis. PCR analysis of clinical samples from patients with these conditions determined that most *Toxoplasma* isolates were type I, type IV, or novel types (69). Outbreaks in Canada and Brazil characterized by severe ocular toxoplasmosis were caused by *Toxoplasma* type I strains (16).

Toxoplasma infections can be diagnosed by serological assays examining the antibody response towards the infection. Identification of *Toxoplasma* oocysts can be identified in the environment using conventional microscopy; however, one must consider the limitation to this approach: (1) the small number of parasites present in environmental samples; and (2) the oocysts are indistinguishable morphologically from other coccidians. *Toxoplasma* oocysts have been isolated from mussels, which serve as paratenic hosts assimilating and concentrating oocysts. *Toxoplasma* oocysts can be identified from water samples using the current USEPA method for concentration of *Cryptosporidium* (54). Centrifugation and flocculation procedures using aluminum sulfate and ferric sulfate can also concentrate *Toxoplasma* oocysts. Sporulated oocysts were recovered more efficiently using aluminum sulfate and unsporulated oocysts could be better recovered using ferric sulfate (63). A TaqMan PCR assay was developed to detect the ssrRNA. Infectious *Toxoplasma* oocysts were detected up to 21 days in mussels as confirmed using the mouse bioassay (8).

Molecular Detection. Most PCR assays used for *Toxoplasma* identification use primers targeting the B1 gene. It is a 35-fold repetitive gene that is highly specific and conserved among strains of *Toxoplasma* (18). A PCR-enzyme immunoassay oligoprobe was developed to detect *Toxoplasma* oocysts. The PCR was directed towards the amplification of the B1 gene. Avidin coated plates were used to capture the biotin-labeled PCR amplicons, and an internal, FITC labeled oligoprobe was allowed to hybridized to the denatured and bound amplicon. The bound FITC-tagged oligoprobe was detected using anti-FITC antibodies tagged with horse radish peroxidase. This assay could detect 50 oocysts in a clean preparation (114). Other assays have focused on the sensitivity of the PCR assay. Jalal designed primers for B1 gene PCR amplification with a sensitivity of 2 parasites/sample (55).

The freeze thaw procedure in Tris-EDTA buffer and proteinase K digestion has been used to break open *Toxoplasma* oocysts. This is followed by DNA extraction using the QIAamp DNA minikit. Again, the B1 gene was used in a real-time PCR using an Icycler device (Bio Rad). Sensitivity of the real-time PCR using experimentally spiked deionized and public drinking water samples was of 1 and 10 oocysts. However, the sensitivity of the PCR was reduced for raw surface waters as only 20%, and 50% were positive by real-time PCR for samples spiked with 100 and 1000 oocysts, respectively (133).

Oocyst heating at 100°C for 40 min in TE buffer followed with 9 freeze-thaw cycles and proteinase K digestion at 56°C overnight has been used to break open the oocysts and provide a template for PCR. DNA was subsequently extracted from the broken oocysts using the phenol: chloroform: isoamyl alcohol procedure. DNA amplification using the18S-rRNA gene designed by McPherson and Gajadhar (77) had a theoretical detection limit of 0.1 oocyst if the preparation included oocyst concentration using aluminum sulfate flocculation (63). TaqMan PCR assays were done using B1 and ssrRNA genes to study experimentally inoculated mussels (8). Real-time PCR was more sensitive to nested PCR using B1 and bradyzoites specific genes. LC-PCR also had the advantage to quantify parasites present in serum and peripheral blood mononuclear cells (23). Other targets were used for DNA amplification. These included a 529 bp sequence present at 300 copies in the parasite genome. Using this target sequence, real-time PCR had a tenfold higher sensitivity compared to PCR targeting the 35 copy B1 gene (52). Mobile genetic elements (MGE), which has 100–500 copies/cell was also used for *Toxoplasma* identification followed by RFLP analysis (128). Other single copy genes SAG1-4 and GRA4 genes have been used as targets for *Toxoplasma* characterization and identification.

Characterization of *Toxoplasma* isolates was achieved using PCR-amplified products digested with 13 restriction enzymes and determined the genetic relationship among *Toxoplasma* isolates and other coccidia (17). RFLP-PCR, random-amplified polymorphic DNA (RAPD), PCR, and sequencing has allowed for genotyping analysis (3, 14, 47). Sequencing of DNA polymerase and the *gra6* genes have also been used to determine strain types (33).

MICROSPORIDIA

Parasite Description and Identification. Microsporidia belongs to the phylum *Microspora*. It contains more than 1000 species and infects a wide range of hosts. Five genera have been implicated in human illness: *Encephalitozoon, Enterocitozoon, Septata, Pleistophora,* and *Vitaforma*. Originally called *S. intestinalis*, it has been reclassified as *Encephalitozoon* and *Nosema cornea* as *Vitaforma cornea*. *E. bieneusi* causes persistant diarrhea in the immunocompromised, where it is found frequently in the feces.

Microsporidia are obligate intracellular organisms that form highly resistant spores of piriform or ovid shape ranging from 1 to 2 μm in diameter. Spores are ingested along with contaminated water or foods, or inhaled. The spore extrudes its polar filament and injects the sporoplasm (infectious spore material). The sporoplasm undergoes merogony forming multiple primordial forms. Sporogony follows, forming dividing sporonts, which form the sporoblast. They then develop into mature spores. Spores are excreted along with the feces, urine, or respiratory secretions. *E. bieneusi* has been reported in mixed infections with *Cryptosporidium*.

Microsporidial spores can be identified microscopically using Calcofluor white, or modified trichrome with Chromotrope 2R stains. The use of other stains has been reported. Identification of the spores is difficult because of the small size of the spores, which require a well-trained microscopist. Molecular assays have played a significant role, particularly in the identification of particular Microsporidia (78). The various species of Microsporidia can be acquired via various routes. The species acquired by ingestion of contaminated foods or water are *E. bieneusi* and *E. intestinalis*. The latter can be successfully grown *in vitro*.

Molecular Detection. Primers were designed to amplify the SSU rRNA gene of *E. bieneusi* in fecal and biopsy samples, either by PCR or *in situ* hybridization (21). PCR protocols call for rupture of Microsporidia spores using glass beads and overnight digestion with proteinase K to release the template. This is followed by DNA extraction using commercial kits and PCR amplification using SSU rRNA gene (24). PCR protocols seem to work with fresh as well as formalized specimens. Other PCR primers have been developed for detecting the various Microsporidia species.

To confirm the identity of the amplified PCR products from the SSU rRNA gene, restriction endoucleases *HaeIII* and *Pst I* have been used to distinguish between *E. bieneusi* and *E. intestinalis* (34); however, these restriction enzymes do not differentiate *E. intestinalis* from *E. cuniculi*. Primers were also designed to amplify conserved segments of the SSU rRNA of these two Microsporidia.

Results were confirmed by standard staining methods and immunofluorescence assay specific for *E. intestinalis* (89). Orlandi evaluated the filter based protocol described for *Cyclospora* and *Cryptosporidium* in PCR detection of *E. intestinalis*. The PCR assay could identify as few as 10–50 *E. intestinalis* spores (92). PCR amplification followed by *Hinf1* endonuclease restriction could identify *E. intestinalis* from clinical specimens (100). Examination of a fragment of the ITS region suggests the presence of genetically distinct strains of *E. bieneusi* (104). Detection of Microsporidia spores in environmental samples was evaluated using immunomagnetic separation followed by PCR and with the concentration of the Microsporidia spores by immunomagnetic separation, PCR could detect as few as 10 spores per 100 L of tap water (117). Real-time quantitative PCR has been developed using the polar tube protein gene 2 of *E. intestinalis*, which could be used to quantify the number of spores produced *in vitro*, or to determine the effect of inactivation procedures (138). Real-time PCR using commercial DNA isolation kits and an automated MagNA Pure LC instrument could identify microsporidia. The sensitivity of the PCR was 100–10,000 spores/ml of feces (145). PCR followed by sequencing of the ITS fragment of various Microsporidia isolates from human and animal origin has demonstrated high variability among isolates, and that few of the isolates from animal origin may be of public health relevance (122).

HELMINTH INFECTIONS

There are three groups of helminthic parasites and each group is significantly relevant to public health. Cestodes or flatworms associated with foodborne outbreaks include *Taenia solium* and *T. saginata, Diphylobrotrium latum, Echinococcus granulosus,* and *E. multilocularis*. Identification of the parasites can be done by observation of the larval (or cystic) stages of the parasite in meat. Nematodes or round worms have also been identified in meats as larval forms (*Trichinella spiralis, Anisakis*) and the eggs can be present in fresh produce or in contaminated water (*Ascaris lumbricoides, Toxocara canis, Capillaria, Gnatostoma,* and *Angiostrongilus*). Flukes or trematodes can also be acquired by ingestion of the cystic forms in fish, crabs, or shellfish (*Paragonimus westermani, Heterophyes,* and *Nanophyetus*). In most instances, these parasites are ingested when foods are eaten fresh or raw (not frozen). Another fluke (*Fasciola* or *Fasciolopsis*) can be acquired by ingestion of raw vegetables containing the metacercaria stages. These parasites can be isolated from meats and produce (39, 75, 112, 129, 144). Molecular assays for detection of these parasites are currently not done as routine procedure; however, laboratories that do molecular epidemiology have developed assays to describe the genotypes most commonly isolated in certain animal species. Most of these parasites have been reported in developing areas and countries, where Good Agricultural Practices (GAPs) are not established. Some of these parasites can be acquired by ingestion of game meats, raw fish, or shellfish. Most of these parasites can be inactivated when frozen.

VIABILITY ASSAYS

The viability of cysts or oocysts has been examined by using vital dyes. *In vitro* cultivation has been successful for some genotypes and species of *Cryptosporidium* and *Toxoplasma*. Most of the microsporidian spores can be propagated *in vitro* except for *E. bieneusi,* which is the most commonly identified microsporidia in humans and associated with gastrointestinal illness (25, 27, 28). To date, there is no effective *in vitro* cultivation assay for *Cyclospora.* *Giardia* can be excysted and grown in TYI-S-33 media; however, a large number of cysts are required for successful propagation. To date, some *Giardia* assemblages or genotypes cannot be propagated *in vitro*. Animal models that could be used to determine infectivity and viability are limited to *Cryptosporidium* and *Toxoplasma* (108). *Cryptosporidium* can be propagated using neonate calves or mice (110). However, *C. hominis*, which is anthroponotic, does not infect these animals. Gnotobiotic pigs have been used to propagate *C. hominis*, but they are not a practical animal model for inactivation and viability studies. No animal model is currently available for *Cyclospora* (31), nor has the disease been reproduced with this agent in healthy, human volunteers. *Toxoplasma* tachyzoites can be propagated using the MRC-5 cell line and most other fibroblast cell lines. It can infect cats, mice, and chickens. Whether this infectivity is selective to certain genotypes is currently not known.

As evident from this discussion, there are several viability assays available for some but not all foodborne parasites. However, how practical and cost-effective are these assays in assessing effectiveness of certain processes for eliminating these pathogens, or reducing their load in food? Obviously, there is a need for developing methods to discern the effectiveness of certain food processes for eradicating or reducing these parasites. Reverse transcriptase PCR developed as "real-time" detection of these foodborne parasites might prove to be an important tool, in the future, for this endeavor.

CONCLUSIONS

Parasites and the intestinal diseases they cause are more frequently being associated with consumption of raw vegetables, fruits, and nonfiltered water. These outbreaks are most often associated with produce imported from areas, where these parasites are endemic. Importation of foods is now necessary in order to satisfy the consumer demands for certain commodities, especially fresh fruits and vegetables. Incentives for importation of foods include the cost of production for particular crops throughout the year. The fast transportation of fresh produce from the farm to the consumers has favored the survival of these pathogens on these commodities. Due to advances in medicine, the U.S. demography has also changed. The elderly, immunocompromised individuals and children are at higher risks of acquiring and having a more severe illness. Most foodborne outbreaks are considered to be bacterial or viral. To determine the etiology of an outbreak may take many days, by which time samples of the implicated product may not be available for investigation. Because of this factor, food scientists and the medical community need to be aware that parasites are significant agents for foodborne outbreaks and routine examination of clinical samples may miss the identification of parasites, particularly with *Cyclospora* and *Cryptosporidium*. Molecular assays have been very helpful in foodborne outbreak investigations, but they have also demonstrated the current limitations in food parasitology. Parasite isolation and recovery procedures are crucial to have a very accurate and specific diagnosis. This is particularly important, since parasites are generally inert in the environment and enrichment procedures necessary for bacterial contaminants are not an option in the detection of these pathogens in foods. With further development and refinement, PCR will prove to be an important tool in surveillance of foods for these emerging human parasites.

REFERENCES

1. Adam, R.D. 2001. Biology of *Giardia lamblia*. Clin. Microbiol. Rev. **14**:447–475.
2. Adam, R.D., Y.R. Ortega, R.H. Gilman, and C.R. Sterling. 2000. Intervening transcribed spacer region 1 variability in *Cyclospora cayetanensis*. J. Clin. Microbiol. **38**:2339–2343.
3. Ajzenberg, D., A.L. Banuls, M. Tibayrenc, and M.L. Darde. 2002. Microsatellite analysis of *Toxoplasma gondii* shows considerable polymorphism structured into two main clonal groups. Int. J. Parasitol. **32**:27–38.
4. Alves, M., O. Matos, and F. Antunes. 2003. Microsatellite analysis of *Cryptosporidium hominis* and *C. parvum* in Portugal: A preliminary study. J. Eukaryot. Microbiol. **50 Suppl**:529–530.
5. Alves, M., L. Xiao, I. Sulaiman, A.A. Lal, O. Matos, and F. Antunes. 2003. Subgenotype analysis of *Cryptosporidium* isolates from humans, cattle, and zoo ruminants in Portugal. J. Clin. Microbiol. **41**:2744–2747.
6. Amar, C.F., P.H. Dear, and J. McLauchlin. 2004. Detection and identification by real-time PCR/RFLP analyses of *Cryptosporidium* species from human faeces. Lett. Appl. Microbiol. **38**:217–222.
7. Amar, C.F., C. East, E. Maclure, J. McLauchlin, C. Jenkins, P. Duncanson, and D.R.A. Wareing. 2004. Blinded application of microscopy, bacteriological culture, immunoassays and PCR to detect gastrointestinal pathogens from faecal samples of patients with community-acquired diarrhoea. Eur. J. Clin. Microbiol. Infect. Dis. **23**:529–534.
8. Arkush, K.D., M.A. Miller, C.M. Leutenegger, I.A. Gardner, A.E. Packham, A.R. Heckeroth, A.M. Tenter, B.C. Barr, and P.A. Conrad. 2003. Molecular and bioassay-based detection of *Toxoplasma gondii* oocyst uptake by mussels (*Mytilus galloprovincialis*). Int. J. Parasit. **33**:1087–1097.
9. Arrowood, M.J. and C.R. Sterling. 1989. Comparison of conventional staining methods and monoclonal antibody-based methods for *Cryptosporidium* oocyst detection. J. Clin. Microbiol. **27**:1490–1495.
10. Baeumner, A.J., M.C. Humiston, R.A. Montagna, and R.A. Durst. 2001. Detection of viable oocysts of *Cryptosporidium parvum* following nucleic acid sequence based amplification. Anal. Chem. **73**:1176–1180.

11. Baeumner, A.J., J. Pretz, and S. Fang. 2004. A universal nucleic acid sequence biosensor with nanomolar detection limits. Anal. Chem. **76**:888–894.

12. Bahia-Oliveira, L.M., J.L. Jones, J. zevedo-Silva, C.C. Alves, F. Orefice, and D.G. Addiss. 2003. Highly endemic, waterborne toxoplasmosis in north Rio de Janeiro state, Brazil. Emerg. Infect. Dis. **9**:55–62.

13. Bier, J.W. 1991. Isolation of parasites on fruits and vegetables. Southeast Asian J. Trop. Med. Public Health **22 Suppl**:144–145.

14. Bohne, W., U. Gross, D.J. Ferguson, and J. Heesemann. 1995. Cloning and characterization of a bradyzoite-specifically expressed gene (*hsp30/bag1*) of *Toxoplasma gondii*, related to genes encoding small heat-shock proteins of plants. Mol. Microbiol. **16**:1221–1230.

15. Bonnin, A., M.N. Fourmaux, J.F. Dubremetz, R.G. Nelson, P. Gobet, G. Harly, M. Buisson, D. Puygauthier-Toubas, G. Gabriel-Pospisil, M. Naciri, and P. Camerlynck. 1996. Genotyping human and bovine isolates of *Cryptosporidium parvum* by polymerase chain reaction-restriction fragment length polymorphism analysis of a repetitive DNA sequence. FEMS Microbiol. Lett. **137**:207–211.

16. Boothroyd, J.C. and M.E. Grigg. 2002. Population biology of *Toxoplasma gondii* and its relevance to human infection: Do different strains cause different disease? Curr. Opin. Microbiol. **5**:438–442.

17. Brindley, P.J., R.T. Gazzinelli, E.Y. Denkers, S.W. Davis, J.P. Dubey, R. Belfort, Jr., M.C. Martins, C. Silveira, L. Jamra, A.P. Waters and A. Sher. 1993. Differentiation of *Toxoplasma gondii* from closely related coccidia by riboprint analysis and a surface antigen gene polymerase chain reaction. Am. J. Trop. Med. Hyg. **48**:447–456.

18. Burg, J. L., C. M. Grover, P. Pouletty, and J. C. Boothroyd. 1989. Direct and sensitive detection of a pathogenic protozoan, *Toxoplasma gondii*, by polymerase chain reaction. J. Clin. Microbiol. **27**:1787–1792.

19. Caccio, S., W. Homan, R. Camilli, G. Traldi, T. Kortbeek, and E. Pozio. 2000. A microsatellite marker reveals population heterogeneity within human and animal genotypes of cryptosporidium parvum. Parasitol. **120**:237–244.

20. Caccio, S., F. Spano, and E. Pozio. 2001. Large sequence variation at two microsatellite loci among zoonotic (genotype C) isolates of *Cryptosporidium parvum*. Int. J. Parasitol. **31**:1082–1086.

21. Carville, A., K. Mansfield, G. Widmer, A. Lackner, D. Kotler, P. Wiest, T. Gumbo, S. Sarbah, and S. Tzipori. 1997. Development and application of genetic probes for detection of *Enterocytozoon bieneusi* in formalin-fixed stools and in intestinal biopsy specimens from infected patients. Clin. Diagn. Lab Immunol. **4**:405–408.

22. Chrisp, C.E. and M. LeGendre. 1994. Similarities and differences between DNA of *Cryptosporidium parvum* and *C. wrairi* detected by the polymerase chain reaction. Folia Parasitol. (Praha) **41**:97–100.

23. Contini, C., S. Seraceni, R. Cultrera, C. Incorvaia, A. Sebastiani, and S. Picot. 2005. Evaluation of a Real-time PCR-based assay using the lightcycler system for detection of *Toxoplasma gondii* bradyzoite genes in blood specimens from patients with toxoplasmic retinochoroiditis. Int. J. Parasitol. **35**:275–283.

24. Da Silva, A.J., F.J. Bornay-Llinares, del Aguila de la Puente Cdel, H. Moura, J.M. Peralta, I. Sobottka, D.A. Schwartz, G.S. Visvesvara, S.B. Slemenda, and N.J. Pieniazek. 1997. Diagnosis of *Enterocytozoon bieneusi* (microsporidia) infections by polymerase chain reaction in stool samples using primers based on the region coding for small-subunit ribosomal RNA. Arch. Pathol. Lab Med. **121**:874–879.

25. Del, A.C., H. Moura, S. Fenoy, R. Navajas, R. Lopez-Velez, L. Li, L. Xiao, G. J. Leitch, S.A. Da, N.J. Pieniazek, A.A. Lal, and G.S. Visvesvara. 2001. In vitro cul-

ture, ultrastructure, antigenic, and molecular characterization of *Encephalitozoon cuniculi* isolated from urine and sputum samples from a Spanish patient with AIDS. J. Clin. Microbiol. **39**:1105–1108.

26. Di Giovanni, G.D. and M.W. Lechevallier. 2005. Quantitative-PCR assessment of *Cryptosporidium parvum* cell culture infection. Appl. Environ. Microbiol. **71**:1495–1500.

27. Didier, E.S., L.B. Rogers, J.M. Orenstein, M.D. Baker, C.R. Vossbrinck, G.T. van, R. Hartskeerl, R. Soave, and L.M. Beaudet. 1996. Characterization of *Encephalitozoon* (*Septata*) *intestinalis* isolates cultured from nasal mucosa and bronchoalveolar lavage fluids of two AIDS patients. J. Eukaryot. Microbiol. **43**:34–43.

28. Doultree, J.C., A.L. Maerz, N.J. Ryan, R.W. Baird, E. Wright, S.M. Crowe, and J.A. Marshall. 1995. In vitro growth of the microsporidian *Septata intestinalis* from an AIDS patient with disseminated illness. J. Clin. Microbiol. **33**:463–470.

29. Dubey, J.P. 2004. Toxoplasmosis -a waterborne zoonosis. Vet. Parasitol. **126**:57–72.

30. Eberhard, M.L., A.J. da Silva, B.G. Lilley, and N.J. Pieniazek. 1999. Morphologic and molecular characterization of new *Cyclospora* species from Ethiopian monkeys: *C. cercopitheci* sp.n., *C. colobi* sp.n., and *C. papionis* sp.n. Emerg. Infect. Dis. **5**:651–658.

31. Eberhard, M.L., Y.R. Ortega, D.E. Hanes, E.K. Nace, R.Q. Do, M.G. Robl, K.Y. Won, C. Gavidia, N.L. Sass, K. Mansfield, A. Gozalo, J. Griffiths, R. Gilman, C.R. Sterling, and M.J. Arrowood. 2000. Attempts to establish experimental *Cyclospora cayetanensis* infection in laboratory animals. J. Parasitol. **86**:577–582.

32. Fayer, R., J.M. Trout, E.J. Lewis, M. Santin, L. Zhou, A.A. Lal, and L. Xiao. 2003. Contamination of Atlantic coast commercial shellfish with *Cryptosporidium*. Parasitol. Res. **89**:141–145.

33. Fazaeli, A., P.E. Carter, M.L. Darde, and T.H. Pennington. 2000. Molecular typing of *Toxoplasma gondii* strains by GRA6 gene sequence analysis. Int. J. Parasitol. **30**:637–642.

34. Fedorko, D.P., N.A. Nelson, and C.P. Cartwright. 1995. Identification of microsporidia in stool specimens by using PCR and restriction endonucleases. J. Clin. Microbiol. **33**:1739–1741.

35. Feng, X., S.M. Rich, D. Akiyoshi, J.K. Tumwine, A. Kekitiinwa, N. Nabukeera, S. Tzipori, and G. Widmer. 2000. Extensive polymorphism in *Cryptosporidium parvum* identified by multilocus microsatellite analysis. Appl. Environ. Microbiol. **66**:3344–3349.

36. Filkorn, R., A. Wiedenmann, and K. Botzenhart. 1994. Selective detection of viable *Cryptosporidium* oocysts by PCR. Zentralbl. Hyg. Umweltmed. **195**:489–494.

37. Fontaine, M. and E. Guillot. 2003. Study of 18S rRNA and rDNA stability by real-time RT-PCR in heat-inactivated *Cryptosporidium parvum* oocysts. FEMS Microbiol. Lett. **226**:237–243.

38. Fuentes, I., J.M. Rubio, C. Ramirez, and J. Alvar. 2001. Genotypic characterization of *Toxoplasma gondii* strains associated with human toxoplasmosis in Spain: Direct analysis from clinical samples. J. Clin. Microbiol. **39**:1566–1570.

39. Gajadhar, A.A. and H.R. Gamble. 2000. Historical perspectives and current global challenges of *Trichinella* and trichinellosis. Vet. Parasitol. **93**:183–189.

40. Garcia, L.S. and W.L. Current. 1989. Cryptosporidiosis: Clinical features and diagnosis. Crit Rev. Clin. Lab Sci. **27**:439–460.

41. Gasser, R.B., Y.G.A. El-Osta, S. Prepens, and R.M. Chalmers. 2004. An improved 'cold SSCP' method for the genotypic and subgenotypic characterization of *Cryptosporidium*. Mol. Cell. Probes **18**:329–332.

42. Gasser, R.B., X.Q. Zhu, S. Caccio, R. Chalmers, G. Widmer, U.M. Morgan, R.C.A. Thompson, E. Pozio, and G.F. Browning. 2001. Genotyping *Cryptosporidium*

parvum by single-strand conformation polymorphism analysis of ribosomal and heat shock gene regions. Electrophor. **22**:433–437.

43. Glaberman, S., J.E. Moore, C.J. Lowery, R.M. Chalmers, I. Sulaiman, K. Elwin, P.J. Rooney, B.C. Millar, J.S. Dooley, A.A. Lal, and L. Xiao. 2002. Three drinking-water-associated cryptosporidiosis outbreaks, Northern Ireland. Emerg. Infect. Dis. **8**:631–633.

44. Gomez-Couso, H., F. Freire-Santos, C.F.L. Amar, K.A. Grant, K. Williamson, M.E. res-Mazas, and J. McLauchlin. 2004. Detection of *Cryptosporidium* and *Giardia* in molluscan shellfish by multiplexed nested-PCR. Int. J. Food Microbiol. **91**:279–288.

45. Gomez-Couso, H., F. Freire-Santos, J. Martinez-Urtaza, O. Garcia-Martin, and M.E. res-Mazas. 2003. Contamination of bivalve molluscs by *Cryptosporidium* oocysts: The need for new quality control standards. Int. J. Food Microbiol. **87**:97–105.

46. Graczyk, T.K., M.R. Cranfield, and R. Fayer. 1996. Evaluation of commercial enzyme immunoassay (EIA) and immunofluorescent antibody (FA) test kits for detection of *Cryptosporidium* oocysts of species other than *Cryptosporidium parvum*. Am. J. Trop. Med. Hyg. **54**:274–279.

47. Guo, Z.G., U. Gross, and A.M. Johnson. 1997. *Toxoplasma gondii* virulence markers identified by random amplified polymorphic DNA polymerase chain reaction. Parasitol. Res. **83**:458–463.

48. Hallier-Soulier, S. and E. Guillot. 2003. An immunomagnetic separation-reverse transcription polymerase chain reaction (IMS-RT-PCR) test for sensitive and rapid detection of viable waterborne *Cryptosporidium parvum*. Environ. Microbiol. **5**:592–598.

49. Higgins, J.A., R. Fayer, J.M. Trout, L.H. Xiao, A.A. Lal, S. Kerby, and M.C. Jenkins. 2001. Real-time PCR for the detection of *Cryptosporidium parvum*. J. Microbiol. Methods **47**:323–337.

50. Ho, A.Y., A.S. Lopez, M.G. Eberhart, R. Levenson, B.S. Finkel, A.J. Da Silva, J. M. Roberts, P.A. Orlandi, C.C. Johnson, and B.L. Herwaldt. 2002. Outbreak of cyclosporiasis associated with imported raspberries, Philadelphia, Pennsylvania, 2000. Emerg. Infect. Dis. **8**:783–788.

51. Homan, W.L. and T.G. Mank. 2001. Human giardiasis: Genotype linked differences in clinical symptomatology. Int. J. Parasitol. **31**:822–826.

52. Homan, W.L., M. Vercammen, B.J. De, and H. Verschueren. 2000. Identification of a 200- to 300-fold repetitive 529 bp DNA fragment in *Toxoplasma gondii*, and its use for diagnostic and quantitative PCR. Int. J. Parasitol. **30**:69–75.

53. Howe, D.K. and L.D. Sibley. 1995. *Toxoplasma gondii* comprises three clonal lineages: Correlation of parasite genotype with human disease. J. Infect. Dis. **172**:1561–1566.

54. Isaac-Renton, J., W.R. Bowie, A. King, G.S. Irwin, C.S. Ong, C.P. Fung, M.O. Shokeir, and J.P. Dubey. 1998. Detection of *Toxoplasma gondii* oocysts in drinking water. Appl. Environ. Microbiol. **64**:2278–2280.

55. Jalal, S., C.E. Nord, M. Lappalainen, and B. Evengard. 2004. Rapid and sensitive diagnosis of *Toxoplasma gondii* infections by PCR. Clin. Microbiol. Infect. **10**:937–939.

56. Jenkins, M.C., J. Trout, M.S. Abrahamsen, C.A. Lancto, J. Higgins, and R. Fayer. 2000. Estimating viability of *Cryptosporidium parvum* oocysts using reverse transcriptase-polymerase chain reaction (RT-PCR) directed at mRNA encoding amyloglucosidase. J. Microbiol. Methods **43**:97–106.

57. Jiang, J.L. and L.H. Xiao. 2003. An evaluation of molecular diagnostic tools for the detection and differentiation of human-pathogenic *Cryptosporidium spp.* J. Euk. Microbiol. **50**:542–547.

58. Jinneman, K.C., J.H. Wetherington, W.E. Hill, A.M. Adams, J.M. Johnson, B.J. Tenge, N.L. Dang, R.L. Manger, and M.M. Wekell. 1998. Template preparation for PCR and RFLP of amplification products for the detection and identification of *Cyclospora spp.* and *Eimeria spp.* oocysts directly from raspberries. J. Food Prot. **61**:1497–1503.

59. Jinneman, K.C., J.H. Wetherington, W.E. Hill, C.J. Omiescinski, A.M. Adams, J.M. Johnson, B.J. Tenge, N.L. Dang, and M.M. Wekell. 1999. An oligonucleotide-ligation assay for the differentiation between *Cyclospora* and *Eimeria spp.* polymerase chain reaction amplification products. J. Food Prot. **62**:682–685.

60. Johnston, S.P., M.M. Ballard, M.J. Beach, L. Causer, and P.P. Wilkins. 2003. Evaluation of three commercial assays for detection of *Giardia* and *Cryptosporidium* organisms in fecal specimens. J. Clin. Microbiol. **41**:623–626.

61. Kappus, K.D., R.G. Lundgren, Jr., D.D. Juranek, J.M. Roberts, and H.C. Spencer. 1994. Intestinal parasitism in the United States: Update on a continuing problem. Am. J. Trop. Med. Hyg. **50**:705–713.

62. Kimbell, L.M., III, D.L. Miller, W. Chavez, and N. Altman. 1999. Molecular analysis of the 18S rRNA gene of *Cryptosporidium serpentis* in a wild-caught corn snake (Elaphe guttata guttata) and a five-species restriction fragment length polymorphism-based assay that can additionally discern *C. parvum* from *C. wrairi*. Appl. Environ. Microbiol. **65**:5345–5349.

63. Kourenti, C., A. Heckeroth, A. Tenter, and P. Karanis. 2003. Development and application of different methods for the detection of *Toxoplasma gondii* in water. Appl. Environ. Microbiol. **69**:102–106.

64. Kozwich, D., K.A. Johansen, K. Landau, C.A. Roehl, S. Woronoff, and P.A. Roehl. 2000. Development of a novel, rapid integrated *Cryptosporidium parvum* detection assay. Appl. Environ. Microbiol. **66**:2711–2717.

65. Laxer, M.A., B.K. Timblin, and R.J. Patel. 1991. DNA sequences for the specific detection of *Cryptosporidium parvum* by the polymerase chain reaction. Am. J. Trop. Med. Hyg. **45**:688–694.

66. Learmonth, J.J., G. Ionas, A.B. Pita, and R.S. Cowie. 2003. Identification and genetic characterisation of *Giardia* and *Cryptosporidium* strains in humans and dairy cattle in the Waikato Region of New Zealand. Water Sci. Technol. **47**:21–26.

67. Lechevallier, M.W., G.D. Di Giovanni, J.L. Clancy, Z. Bukhari, S. Bukhari, J.S. Rosen, J. Sobrinho, and M.M. Frey. 2003. Comparison of method 1623 and cell culture-PCR for detection of *Cryptosporidium spp.* in source waters. Appl. Environ. Microbiol. **69**:971–979.

68. Lechevallier, M.W., W.D. Norton, and R.G. Lee. 1991. *Giardia* and *Cryptosporidium spp.* in filtered drinking water supplies. Appl. Environ. Microbiol. **57**:2617–2621.

69. Lehmann, T., C.R. Blackston, S.F. Parmley, J.S. Remington, and J.P. Dubey. 2000. Strain typing of *Toxoplasma gondii*: Comparison of antigen-coding and housekeeping genes. J. Parasitol. **86**:960–971.

70. Leng, X., D.A. Mosier, and R.D. Oberst. 1996. Differentiation of *Cryptosporidium parvum*, *C. muris*, and *C. baileyi* by PCR-RFLP analysis of the 18S rRNA gene. Vet. Parasitol. **62**:1–7.

71. Leoni, F., C.I. Gallimore, J. Green, and J. McLauchlin. 2003. A rapid method for identifying diversity within PCR amplicons using a heteroduplex mobility assay and

synthetic polynucleotides: Application to characterisation of dsRNA elements associated with *Cryptosporidium*. J. Microbiol. Methods **54**:95–103.

72. Lindquist, H.D., M. Ware, R.E. Stetler, L. Wymer, and F.W. Schaefer, III. 2001. A comparison of four fluorescent antibody-based methods for purifying, detecting, and confirming *Cryptosporidium parvum* in surface waters. J. Parasitol. **87**:1124–1131.

73. Lopez, F.A., J. Manglicmot, T.M. Schmidt, C. Yeh, H.V. Smith, and D.A. Relman. 1999. Molecular characterization of *Cyclospora*-like organisms from baboons. J. Infect. Dis. **179**:670–676.

74. Lowery, C.J., J.E. Moore, B.C. Millar, D.P. Burke, K.A. McCorry, E. Crothers, and J.S. Dooley. 2000. Detection and speciation of *Cryptosporidium spp.* in environmental water samples by immunomagnetic separation, PCR and endonuclease restriction. J. Med. Microbiol. **49**:779–785.

75. Lun, Z.R., R.B. Gasser, D.H. Lai, A.X. Li, X.Q. Zhu, X.B. Yu, and Y.Y. Fang. 2005. Clonorchiasis: A key foodborne zoonosis in China. Lancet Infect. Dis. **5**:31–41.

76. MacKenzie, W.R., W.L. Schell, K.A. Blair, D.G. Addiss, D.E. Peterson, N.J. Hoxie, J.J. Kazmierczak, and J.P. Davis. 1995. Massive outbreak of waterborne cryptosporidium infection in Milwaukee, Wisconsin: Recurrence of illness and risk of secondary transmission. Clin. Infect. Dis. **21**:57–62.

77. MacPherson, J.M., and A.A. Gajadhar. 1993. Sensitive and specific polymerase chain reaction detection of *Toxoplasma gondii* for veterinary and medical diagnosis. Can. J. Vet. Res. **57**:45–48.

78. Marshall, M.M., D. Naumovitz, Y. Ortega, and C.R. Sterling. 1997. Waterborne protozoan pathogens. Clin. Microbiol. Rev. **10**:67–85.

79. Monis, P.T., and R.H. Andrews. 1998. Molecular epidemiology: Assumptions and limitations of commonly applied methods. Int. J. Parasitol. **28**:981–987.

80. Monis, P.T., R.H. Andrews, G. Mayrhofer, and P.L. Ey. 1999. Molecular systematics of the parasitic protozoan *Giardia intestinalis*. Mol. Biol. Evol. **16**:1135–1144.

81. Monis, P.T., G. Mayrhofer, R.H. Andrews, W.L. Homan, L. Limper, and P.L. Ey. 1996. Molecular genetic analysis of *Giardia intestinalis* isolates at the glutamate dehydrogenase locus. Parasitol **112**:1–12.

82. Morgan, U.M., C.C. Constantine, D.A. Forbes, and R.C. Thompson. 1997. Differentiation between human and animal isolates of *Cryptosporidium parvum* using rDNA sequencing and direct PCR analysis. J. Parasitol. **83**:825–830.

83. Morgan, U.M., C.C. Constantine, P. O'Donoghue, B.P. Meloni, P.A. O'Brien, and R.C. Thompson. 1995. Molecular characterization of *Cryptosporidium* isolates from humans and other animals using random amplified polymorphic DNA analysis. Am. J. Trop. Med. Hyg. **52**:559–564.

84. Morgan, U.M., P. Deplazes, D.A. Forbes, F. Spano, H. Hertzberg, K.D. Sargent, A. Elliot, and R.C. Thompson. 1999. Sequence and PCR-RFLP analysis of the internal transcribed spacers of the rDNA repeat unit in isolates of *Cryptosporidium* from different hosts. Parasitol. **118**:49–58.

85. Moss, D.M. and P.J. Lammie. 1993. Proliferative responsiveness of lymphocytes from *Cryptosporidium parvum*-exposed mice to two separate antigen fractions from oocysts. Am. J. Trop. Med. Hyg. **49**:393–401.

86. Nichols, R.A., B.M. Campbell, and H.V. Smith. 2003. Identification of *Cryptosporidium spp.* oocysts in United Kingdom noncarbonated natural mineral waters and drinking waters by using a modified nested PCR-restriction fragment length polymorphism assay. Appl. Environ. Microbiol. **69**:4183–4189.

87. Nielsen, C.K. and L.A. Ward. 1999. Enhanced detection of *Cryptosporidium parvum* in the acid-fast stain. J. Vet. Diagn. Invest **11**:567–569.

88. Olivier, C., P.S. van De, P.W. Lepp, K. Yoder, and D.A. Relman. 2001. Sequence variability in the first internal transcribed spacer region within and among *Cyclospora* species is consistent with polyparasitism. Int. J. Parasitol. **31**:1475–1487.

89. Ombrouck, C., L. Ciceron, S. Biligui, S. Brown, P. Marechal, G.T. Van, A. Datry, M. Danis, and I. sportes-Livage. 1997. Specific PCR assay for direct detection of intestinal microsporidia *Enterocytozoon bieneusi* and *Encephalitozoon intestinalis* in fecal specimens from human immunodeficiency virus-infected patients. J. Clin. Microbiol. **35**:652–655.

90. Orlandi, P.A., L. Carter, A.M. Brinker, A.J. Da Silva, D.M. Chu, K.A. Lampel, and S.R. Monday. 2003. Targeting single-nucleotide polymorphisms in the 18S rRNA gene to differentiate *Cyclospora* species from *Eimeria* species by multiplex PCR. Appl. Environ. Microbiol. **69**:4806–4813.

91. Orlandi, P.A., D.M. Chu, J.W. Bier, and G.J. Jackson. 2002. Parasites and the Food Supply. Food Technol. **56**:72–81.

92. Orlandi, P.A. and K.A. Lampel. 2000. Extraction-free, filter-based template preparation for rapid and sensitive PCR detection of pathogenic parasitic protozoa. J. Clin. Microbiol. **38**:2271–2277.

93. Ortega, Y.R., R.H. Gilman, and C.R. Sterling. 1994. A new coccidian parasite (*Apicomplexa: Eimeriidae*) from humans. J. Parasitol. **80**:625–629.

94. Ortega, Y.R., R. Nagle, R.H. Gilman, J. Watanabe, J. Miyagui, H. Quispe, P. Kanagusuku, C. Roxas, and C.R. Sterling. 1997. Pathologic and clinical findings in patients with cyclosporiasis and a description of intracellular parasite life-cycle stages. J. Infect. Dis. **176**:1584–1589.

95. Ortega, Y.R., C.R. Roxas, R.H. Gilman, N.J. Miller, L. Cabrera, C. Taquiri, and C.R. Sterling. 1997. Isolation of *Cryptosporidium parvum* and *Cyclospora cayetanensis* from vegetables collected in markets of an endemic region in Peru. Am. J. Trop. Med. Hyg. **57**:683–686.

96. Ortega, Y.R., C.R. Sterling, R.H. Gilman, V.A. Cama, and F. Diaz. 1993. *Cyclospora* species—a new protozoan pathogen of humans. N. Engl. J. Med. **328**:1308–1312.

97. Patel, S., S. Pedraza-Diaz, and J. McLauchlin. 1999. The identification of *Cryptosporidium* species and *Cryptosporidium parvum* directly from whole faeces by analysis of a multiplex PCR of the 18S rRNA gene and by PCR/RFLP of the *Cryptosporidium* outer wall protein (COWP) gene. Int. J. Parasitol. **29**:1241–1247.

98. Peng, M.M., S.R. Meshnick, N.A. Cunliffe, B.D. Thindwa, C.A. Hart, R.L. Broadhead, and L. Xiao. 2003. Molecular epidemiology of cryptosporidiosis in children in Malawi. J. Eukaryot. Microbiol. **50 Suppl**:557–559.

99. Priest, J.W., J.P. Kwon, M.J. Arrowood, and P.J. Lammie. 2000. Cloning of the immunodominant 17-kDa antigen from *Cryptosporidium parvum*. Mol. Biochem. Parasitol. **106**:261–271.

100. Raynaud, L., F. Delbac, V. Broussolle, M. Rabodonirina, V. Girault, M. Wallon, G. Cozon, C. P. Vivares, and F. Peyron. 1998. Identification of *Encephalitozoon intestinalis* in travelers with chronic diarrhea by specific PCR amplification. J. Clin. Microbiol. **36**:37–40.

101. Relman, D.A., T.M. Schmidt, A. Gajadhar, M. Sogin, J. Cross, K. Yoder, O. Sethabutr, and P. Echeverria. 1996. Molecular phylogenetic analysis of *Cyclospora*, the human intestinal pathogen, suggests that it is closely related to *Eimeria* species. J. Infect. Dis. **173**:440–445.

102. Reperant, J.M., M. Naciri, S. Iochmann, M. Tilley, and D.T. Bout. 1994. Major antigens of *Cryptosporidium parvum* recognised by serum antibodies from different infected animal species and man. Vet. Parasitol. **55**:1–13.

103. Rigo, C.R. and R.M. Franco. 2002. [Comparison between the modified Ziehl-Neelsen and Acid-Fast-Trichrome methods for fecal screening of *Cryptosporidium parvum* and Isospora belli]. Rev. Soc. Bras. Med. Trop. **35**:209–214.

104. Rinder, H., S. Katzwinkel-Wladarsch, and T. Loscher. 1997. Evidence for the existence of genetically distinct strains of *Enterocytozoon bieneusi*. Parasitol. Res. **83**:670–672.

105. Robertson, L.J. and B. Gjerde. 2001. Factors affecting recovery efficiency in isolation of *Cryptosporidium* oocysts and *Giardia* cysts from vegetables for standard method development. J. Food Prot. **64**:1799–1805.

106. Robertson, L.J. and B. Gjerde. 2000. Isolation and enumeration of *Giardia* cysts, *Cryptosporidium* oocysts, and *Ascaris* eggs from fruits and vegetables. J. Food Prot. **63**:775–778.

107. Robertson, L.J., B. Gjerde, and A.T. Campbell. 2000. Isolation of *Cyclospora* oocysts from fruits and vegetables using lectin-coated paramagnetic beads. J. Food Prot. **63**:1410–1414.

108. Rochelle, P.A., D.M. Ferguson, T. J. Handojo, L.R. De, M.H. Stewart, and R.L. Wolfe. 1997. An assay combining cell culture with reverse transcriptase PCR to detect and determine the infectivity of waterborne *Cryptosporidium parvum*. Appl. Environ. Microbiol. **63**:2029–2037.

109. Rochelle, P.A., E.M. Jutras, E.R. Atwill, L.R. De, and M.H. Stewart. 1999. Polymorphisms in the beta-tubulin gene of *Cryptosporidium parvum* differentiate between isolates based on animal host but not geographic origin. J. Parasitol. **85**:986–989.

110. Rochelle, P.A., M.M. Marshall, J.R. Mead, A.M. Johnson, D.G. Korich, J. S. Rosen, and L.R. De. 2002. Comparison of in vitro cell culture and a mouse assay for measuring infectivity of *Cryptosporidium parvum*. Appl. Environ. Microbiol. **68**:3809–3817.

111. Rodgers, M.R., D.J. Flanigan, and W. Jakubowski. 1995. Identification of algae which interfere with the detection of *Giardia* cysts and *Cryptosporidium* oocysts and a method for alleviating this interference. Appl. Environ. Microbiol. **61**:3759–3763.

112. Rodriguez-Canul, R., F. rgaez-Rodriguez, G. D. De, S. Villegas-Perez, A. Fraser, P.S. Craig, L. Cob-Galera, and J.L. Dominguez-Alpizar. 2002. *Taenia solium* metacestode viability in infected pork after preparation with salt pickling or cooking methods common in Yucatan, Mexico. J. Food Prot. **65**:666–669.

113. Rose, J.B. and T.R. Slifko. 1999. *Giardia, Cryptosporidium*, and *Cyclospora* and their impact on foods: A review. J. Food Prot. **62**:1059–1070.

114. Schwab, K.J. and J.J. McDevitt. 2003. Development of a PCR-enzyme immunoassay oligoprobe detection method for *Toxoplasma gondii* oocysts, incorporating PCR controls. Appl. Environ. Microbiol. **69**:5819–5825.

115. Simmons, O.D., M.D. Sobsey, F.W. Schaefer, D.S. Francy, R.A. Nally, and C.D. Heaney. 2001. Evaluation of USEPA method 1622 for detection of *Cryptosporidium* oocysts in stream waters. J. Amer. Water Works Assoc. **93**:78–87.

116. Smith, H.V., B.M. Campbell, C.A. Paton, and R.A. Nichols. 2002. Significance of enhanced morphological detection of *Cryptosporidium sp.* oocysts in water concentrates determined by using 4′,6′-diamidino-2-phenylindole and immunofluorescence microscopy. Appl. Environ. Microbiol. **68**:5198–5201.

117. Sorel, N., E. Guillot, M. Thellier, I. Accoceberry, A. Datry, L. Mesnard-Rouiller, and M. Miegeville. 2003. Development of an immunomagnetic separation-polymerase chain reaction (IMS-PCR) assay specific for *Enterocytozoon bieneusi* in water samples. J. Appl. Microbiol. **94**:273–279.

118. Spano, F., L. Putignani, A. Crisanti, P. Sallicandro, U.M. Morgan, S.M. Le Blancq, L. Tchack, S. Tzipori, and G. Widmer. 1998. Multilocus genotypic analysis of *Cryptosporidium parvum* isolates from different hosts and geographical origins. J. Clin. Microbiol. **36**:3255–3259.

119. Straub, T.M., D.S. Daly, S. Wunshel, P.A. Rochelle, R. DeLeon, and D.P. Chandler. 2002. Genotyping *Cryptosporidium parvum* with an *hsp70* single-nucleotide polymorphism microarray. Appl. Environ. Microbiol. **68**:1817–1826.

120. Sturbaum, G.D., C. Reed, P.J. Hoover, B.H. Jost, M.M. Marshall, and C.R. Sterling. 2001. Species-specific, nested PCR-restriction fragment length polymorphism detection of single *Cryptosporidium parvum* oocysts. Appl. Environ. Microbiol. **67**:2665–2668.

121. Sulaiman, I.M., R. Fayer, C. Bern, R.H. Gilman, J.M. Trout, P.M. Schantz, P. Das, A.A. Lal, and L. Xiao. 2003. Triosephosphate isomerase gene characterization and potential zoonotic transmission of *Giardia duodenalis*. Emerg. Infect. Dis. **9**:1444–1452.

122. Sulaiman, I.M., R. Fayer, C. Yang, M. Santin, O. Matos, and L. Xiao. 2004. Molecular characterization of *Enterocytozoon bieneusi* in cattle indicates that only some isolates have zoonotic potential. Parasitol. Res. **92**:328–334.

123. Sulaiman, I.M., J. Jiang, A. Singh, and L. Xiao. 2004. Distribution of *Giardia duodenalis* genotypes and subgenotypes in raw urban wastewater in Milwaukee, Wisconsin. Appl. Environ. Microbiol. **70**:3776–3780.

124. Sulaiman, I.M., A.A. Lal, and L. Xiao. 2001. A population genetic study of the *Cryptosporidium parvum* human genotype parasites. J. Eukaryot. Microbiol. **Suppl**:24S–27S.

125. Sulaiman, I.M., A.A. Lal, and L. Xiao. 2002. Molecular phylogeny and evolutionary relationships of *Cryptosporidium* parasites at the actin locus. J. Parasitol. **88**:388–394.

126. Sulaiman, I.M., U.M. Morgan, R.C. Thompson, A.A. Lal, and L. Xiao. 2000. Phylogenetic relationships of *Cryptosporidium* parasites based on the 70-kilodalton heat shock protein (HSP70) gene. Appl. Environ. Microbiol. **66**:2385–2391.

127. Sulaiman, I.M., L. Xiao, and A.A. Lal. 1999. Evaluation of *Cryptosporidium parvum* genotyping techniques. Appl. Environ. Microbiol. **65**:4431–4435.

128. Terry, R.S., J.E. Smith, P. Duncanson, and G. Hide. 2001. MGE-PCR: A novel approach to the analysis of *Toxoplasma gondii* strain differentiation using mobile genetic elements. Int. J. Parasitol. **31**:155–161.

129. van, H.L., D. Blair, and T. Agatsuma. 1999. Genetic diversity in parthenogenetic triploid *Paragonimus westermani*. Int. J. Parasitol. **29**:1477–1482.

130. van, K.H., P.T. Macechko, S. Wade, S. Schaaf, P.M. Wallis, and S.L. Erlandsen. 2002. Presence of human *Giardia* in domestic, farm and wild animals, and environmental samples suggests a zoonotic potential for giardiasis. Vet. Parasitol. **108**:97–107.

131. van, K.H., P.A. Steimle, D.A. Bulik, R.K. Borowiak, and E.L. Jarroll. 1998. Cloning of two putative *Giardia lamblia* glucosamine 6-phosphate isomerase genes only one of which is transcriptionally activated during encystment. J. Eukaryot. Microbiol. **45**:637–642.

132. Varma, M., J.D. Hester, F.W. Schaefer, M.W. Ware, and H.D.A. Lindquist. 2003. Detection of *Cyclospora cayetanensis* using a quantitative real-time PCR assay. J. Microbiol. Methods **53**:27–36.

133. Villena, I., D. Aubert, P. Gomis, H. Ferte, J.C. Inglard, H. is-Bisiaux, J.M. Dondon, E. Pisano, N. Ortis, and J.M. Pinon. 2004. Evaluation of a strategy for *Toxoplasma gondii* oocyst detection in water. Appl. Environ. Microbiology 70:4035–4039.

134. Wagner-Wiening, C. and P. Kimmig. 1995. Detection of viable *Cryptosporidium parvum* oocysts by PCR. Appl. Environ. Microbiol. 61:4514–4516.

135. Wallis, P.M., D. Matson, M. Jones, and J. Jamieson. 2001. Application of monitoring data for *Giardia* and *Cryptosporidium* to boil water advisories. Risk Anal. 21:1077–1085.

136. Ward, P.I., P. Deplazes, W. Regli, H. Rinder, and A. Mathis. 2002. Detection of eight *Cryptosporidium* genotypes in surface and waste waters in Europe. Parasitol. 124:359–368.

137. Ware, M.W., L. Wymer, H.D.A. Lindquist, and F.W. Schaefer. 2003. Evaluation of an alternative IMS dissociation procedure for use with Method 1622: Detection of *Cryptosporidium* in water. J. Microbiol. Methods 55:575–583.

138. Wasson, K. and P.A. Barry. 2003. Molecular characterization of *Encephalitozoon intestinalis* (*Microspora*) replication kinetics in a murine intestinal cell line. J. Eukaryot. Microbiol. 50:169–174.

139. Weber, R., R.T. Bryan, and D.D. Juranek. 1992. Improved stool concentration procedure for detection of *Cryptosporidium* oocysts in fecal specimens. J. Clin. Microbiol. 30:2869–2873.

140. Webster, K.A., J.D. Pow, M. Giles, J. Catchpole, and M.J. Woodward. 1993. Detection of *Cryptosporidium parvum* using a specific polymerase chain reaction. Vet. Parasitol. 50:35–44.

141. Widmer, G. 1998. Genetic heterogeneity and PCR detection of *Cryptosporidium parvum*. Adv. Parasitol. 40:223–239.

142. Widmer, G., X. Feng, and S. Tanriverdi. 2004. Genotyping of *Cryptosporidium parvum* with microsatellite markers. Methods Mol. Biol. 268:177–187.

143. Widmer, G., E.A. Orbacz, and S. Tzipori. 1999. beta-tubulin mRNA as a marker of *Cryptosporidium parvum* oocyst viability. Appl. Environ. Microbiol. 65:1584–1588.

144. Wilson, M.E., C.A. Lorente, J.E. Allen, and M.L. Eberhard. 2001. Gongylonema infection of the mouth in a resident of Cambridge, Massachusetts. Clin. Infect. Dis. 32:1378–1380.

145. Wolk, D.M., S.K. Schneider, N.L. Wengenack, L.M. Sloan, and J.E. Rosenblatt. 2002. Real-time PCR method for detection of *Encephalitozoon intestinalis* from stool specimens. J. Clin. Microbiol. 40:3922–3928.

146. Xiao, L., A. Singh, J. Limor, T.K. Graczyk, S. Gradus, and A. Lal. 2001. Molecular characterization of *Cryptosporidium* oocysts in samples of raw surface water and wastewater. Appl. Environ. Microbiol. 67:1097–1101.

147. Xiao, L., I.M. Sulaiman, U.M. Ryan, L. Zhou, E.R. Atwill, M.L. Tischler, X. Zhang, R. Fayer, and A. A. Lal. 2002. Host adaptation and host-parasite co-evolution in *Cryptosporidium*: Implications for taxonomy and public health. Int. J. Parasitol. 32:1773–1785.

148. Zu, S.X., G.D. Fang, R. Fayer, and R.L. Guerrant. 1992. Cryptosporidiosis: Pathogenesis and immunology. Parasitol. Today 8:24–27.

Table 7.1. Primers used for the identification of protozoan parasites.

Organism	Target Gene	Sequence	Expected Size (bp)	Reference
Cryptosporidium	18S rRNA	F:AGCTCGTAGTTGGATTTCTG	435	92
		R:TAAGGTGCTGAAGGAGTAAGG		
	SSU-rRNA	F:TTCTAGAGCTAATACATGCG[1]	1,325	146
		R:CCCATTTCCTTCGAAACAGGA[1]		
		F:GGAAGGGTTGTATTTATTAGATAAAG[2]	826–864	48
		R:AAGGAGTAAGGAACAACCTCCA[2]		
	chromosomal	F:CCGAGTTTGATCCAAAAAGTTACGAA[1]	402	
	fragment	R:TAGCTCCTCATATGCCTTATTGAGTA[1]		
		F:GCGAAGATGACCTTTTGATTTG[2]	194	
		R:AGGATTTCTTCTTCTGAGGTTCC[2]		
Cyclospora cayetanensis	18S rRNA	F:TACCCAATGAAAACAGTTT[1]	636	92
		R:CAGGAGAAGCCAAGGTAGG[1]		
		F:CCTTCCGCGCTTCGCTGCGT[2]	294	
		R:CGTCTTCAAACCCCTACTG[2]		
Encephalitozoon intestinalis	18S rRNA	F:TTTCGAGTGTAAAGGAGTCGA	520	92
		R:CCGTCCTCGTTCTCCTGCCCG		
Microsporidia	18S rRNA	F:CACCAGGTTGATTCTGCCTGA[3]	1,300	
		R:TAATGATCCTGCTAATGGTTCTCCAAC[3]		
Enterocytozoon bieneusi	18S rRNA	F:GAAACTTGTCCACTCCTTACG[3]	607	92
		R:CAATGCACCACTCCTGCCATT[3]		
Encephalitozoon cuniculi	18S rRNA	F:ATGAGAAGTGATGTGTGTGCG[3]	549	92
		R:TGCCATGCACTCACAGGCATC[3]		
Encephalitozoon hellem	18S rRNA	F:TGAGAAGTAAGATGTTTAGCA[3]	547	92

Giardia	18S rRNA	R:GTAAAAACACTCTCACACTCA[3]		
		F:GCGCACCAGGAATGTCTTGT	183	130
		R:TCACCTACGGATACCTTGTT		
Toxoplasma gondii	SSU rRNA	F:CCGGTGGTCCTCAGGTGAT	120	8
		R:TGCCACGGTAGTCCAATACAGTA		
		FAM-ATCGCGTTGACTTCGGTCTGCGC-TAMRA[4]		
	B1	F:TCGAAGCTGAGATGCTCAAAGTC	129	8
		R:AATCCACGTCTGGGAAGAACTC		
		FAM-ACCGCGAGATGCACCCGCA-TAMRA[4]		
	B1	F:Biotin-GGAACTGCATCCGTTCATGAG[5]	193	114
		R:TCTTTAAAGCGTTCGTGGTC[5]		
		FITC-GGCGACCAATCTGCGAATACACC[5]		
		FITC-TCGTCAGTGACTGCAACCTATGC[5]		

[1]Nested PCR, external primers
[2]Nested PCR, internal primers
[3]Multiplex PCR primers
[4]TaqMan primer. FAM: 6-carboxyfluorescein, TAMRA, 6-carboxytetramethylrhodamine.
[4]PCR-ELISA. See text or reference 114 for details.

INDEX

16S rDNA, 11–12
Amplicons, non-specific, 28–29
Commercial PCR
  Campylobacter, 63–64
  Escherichia coli, O157:H7, 63–64
    pathogenic, 63–64
  Listeria monocytogenes, 63–64
  Salmonella, 63–64, 65
Controls, 28, 32, 46, 59–60
DNA databases, 11
  NCBI, 14–19
  searches, 14–19
False-negatives, PCR inhibitors, 34–35
False-positives
  live vs. dead cells, 33–34
  non-specific amplicons, 29
  primer-dimers, 29
  viable but nonculturable (VBNC), 34
Interpretation of PCR results, 28–31
Laboratory
  equipment, 57–58
  personnel, 55
  reagents and disposables, 58–59, 60
  set-up, 52–55
  standard operating procedures, 59–60
  standardization, 62
Microarrays, 12
Multiplex PCR, 10, 70, 84
Multiplex PCR, 70, 84
Nucleic acid extraction
  parasites, 120–121
  viruses, 99–100
Nucleic acid sequence-based amplification
    (NASBA), viruses, 102, 107
parvum, 124–125
PCR contamination, 52–55
PCR design, 12–19
PCR detection methods, 6–8
  agarose gels, 6
  ELISA, 6–8, 56
  fluorescence, 8–10, 56–57
PCR detection of
  Bacillus cereus, 71
  Campylobacter, 63–64, 71–72, 85
  Clostridium perfringens, 72–73
  Cryptosporidium parvum, 123–124, 144
  Cyclospora cayetanensis, 126–127, 144
  Escherichia coli O157:H7, 63–64, 85

  Escherichia coli, pathogenic, 63–64,
    73–76, 84–85
  Giardia intestinalis, 128, 145
  hepatitis A virus, 102, 105–106
  Listeria monocytogenes, 63–64, 76–77,
    85
  Microsporidia, 131–132, 144
  noroviruses, 100–101, 103–104
  Salmonella, 63–64, 85, 77–79, 84–85
  Shigella, 79–80
  Staphylococcus aureus, 80–81
  Toxoplasma gondii, 130–131, 145
  Vibrio, pathogenic, 81–83, 85
  Yersinia enterocolitica, 83, 85
PCR inhibitors, 34–35, 46–47
PCR theory, 2–4
Quality control, 59–60
Real-time PCR
  Campylobacter, 63–64, 85
  concept, 8–10
  Escherichia coli O157:H7, 63–64, 85
    pathogenic, 63–64, 85
  false-positives, 10
  hepatitis A virus, 108–109
  Listeria monocytogenes, 63–64, 85
  noroviruses, 108–109
  Salmonella, 63–64, 85
  TaqMan, 8
  Vibrio, pathogenic, 85,
  Yersinia enterocolitica, 85
Reverse-transcriptase PCR, detection of
  bacteria, 33–34, 84–85
  viruses, 101–102
Sample preparation, 41–49
  concentrating microbes, 47–49, 93–99,
    122–123,
  dairy, 43, 96–99
  enrichment, 47–49
  fruits and produce, 45, 96–99
  meat, 43–44
  seafood, 44–45, 94–96
Sensitivity
  false-negatives, 31
  limit of detection, 34–35, 47–49, 70
  PCR inhibitors, 34–35, 46–47
  false-positives, 31
Terminal Restriction Fragment Length
    Polymorphisms, 10–12

Thermocycler
  conventional, 4, 57–58
  gradient, 5
  hot-air capillary, 4–5, 57–58
  real-time PCR, 5–6, 57–58
Thermocyclers, 4–6, 57–58
Trouble-shooting
  false-negatives, 34–35, 46–47
  false-positives, 32–33, 45

  real-time PCR, 10, 29
  sensitivity, 47–49
Validation of PCR, 31, 60–62
Viability
  bacteria, 33–34
  parasites, 133
Virus, concentration, 93–99